AF885088

SAM ALTMAN

The Making of Stargate

First Edition

ISBN: **978-1-967440-05-4**

Published by AIMQWEST Corporation

AIMQWEST BOOKS
AIMQWESTBOOKS.COM

Disclaimer:

This book is a work of fiction. Names, characters, businesses, places, events, locales, and incidents are either the products of the author's imagination or used in a fictitious manner. Any resemblance to actual persons, living or dead, or actual events, institutions, organizations, technologies, or places is purely coincidental and unintended. The views, opinions, and ideas expressed herein do not reflect those of AIMQWEST BOOKS or its affiliates. AIMQWEST BOOKS, its officers, successors, and assigns explicitly disclaim any liability resulting from the reader's interpretation or application of any information contained within this fictional narrative.

TABLE OF CONTENTS

Preface

In every era, there arises a figure whose imagination, conviction, and timing align so perfectly with the moment's technological currents that their impact reshapes not only an industry, but civilization itself. Sam Altman is such a figure for the age of artificial intelligence. While many technologists have helped build the digital infrastructure of the 21st century, few have dared to guide its soul. Fewer still have attempted what Altman has—marrying raw computational power with a sense of philosophical responsibility toward humanity's collective future.

This book, *Sam Altman and the Making of Stargate*, chronicles not just the development of a groundbreaking technological platform, but the emergence of a new human narrative—one in which man and machine must coexist not in tension, but in partnership. It is the story of how a single visionary, through deliberate design and ethical fortitude, constructed more than an AI system—he helped lay the groundwork for a new social contract between intelligence, innovation, and intention.

Stargate is no ordinary product of innovation. It is an infrastructural leap that integrates artificial general intelligence, global information symmetry, and human-centered systems design into one cohesive platform. What makes this story urgent and profoundly compelling is not just the technology itself—but the soul behind its architecture. That soul is Sam Altman.

In tracing the origins of Stargate, we begin with Altman's formative years—his childhood curiosities, academic instincts, and the early coding experiments that hinted at something greater to come. We follow his trajectory through Silicon Valley, where he would eventually rise to lead OpenAI, transforming it from a speculative non-profit into the most strategically positioned AI company in the world.

But this is not a hagiography. This book does not lionize Altman blindly, nor does it ignore the legitimate fears, criticisms, and controversies surrounding artificial intelligence and his role in its proliferation. Instead, it paints a textured, honest portrait of a man grappling with power, risk, and responsibility at a scale few have ever known. Through it all, the consistent thread is Altman's unwavering focus on the long-term—his belief that the future is not a mystery to be feared but a blueprint to be shaped by ethical intelligence.

What follows is both a biography and a technological epic. It is a record of the minds and machines that built Stargate, the politics and promises of its global rollout, and the reverberations it has begun to create across nations, industries, and lives. We include never-before-published insights from engineers, researchers, ethicists, policymakers, and even critics. Their voices shape this narrative, not to dilute Altman's vision, but to highlight the stakes of its realization.

Stargate is more than a tool. It is a threshold. This book is the map of how we arrived at its edge—and what lies beyond.

The journey of building Stargate was never linear. It was a multi-threaded operation woven from technical breakthroughs, existential debates, sleepless code sprints, and high-stakes diplomatic engagements. Its architecture is both technical and philosophical, marrying machine learning with mission intent—its code shaped as much by values as by variables.

Altman's leadership style—calm, calculated, and quietly provocative—enabled him to marshal minds from across disciplines and continents. He did not merely gather engineers and data scientists. He convened philosophers, cognitive scientists, legal theorists, geopolitical strategists, and artists. He understood from the outset that Stargate could not just be engineered; it had to be envisioned. For an intelligence system to serve all of humanity, it would need to be seeded with more than code. It would require conscience.

But vision alone was never enough. Stargate's genesis coincided with an era of mounting global distrust—of institutions, of data monopolies, of AI systems increasingly perceived as opaque and unaccountable. Against this backdrop, Altman insisted on radical transparency. He advocated for alignment not just as a technical problem but as a human commitment—a pledge to tether intelligence to intent, power to principle.

Stargate's development also raised foundational questions that humanity had never truly faced before. What happens when AI knows more about us than we know about ourselves? Can free will persist in a world where predictions become prescriptions? Who gets to write the rules of engagement for machines that reason, adapt, and evolve?

Rather than avoiding these questions, Altman invited them into the design room. This willingness to embrace complexity is what differentiates Stargate from other AI platforms. It was not designed merely to serve, but to elevate. It was not built to outthink humans, but to think with them—to amplify our strengths, compensate for our weaknesses, and anticipate our needs without exploiting them.

This book is organized around that principle: augmentation, not replacement. Each chapter unpacks a layer of this transformation—from the roots of Altman's thinking to the infrastructure of OpenAI, from the birth of the Stargate initiative to the ethical firestorms that challenged its path. Through it all, we see a mind at work not in isolation, but in collaboration—with ideas, with people, and with the future itself.

In reading this story, you may see echoes of your own hopes and fears about technology. That is by design. *Sam Altman and the Making of Stargate* is not a passive biography. It is a mirror held up to this moment in history—an invitation to reflect, to reckon, and to participate.

Because the future is not yet written. And as this story reveals, those who dare to write it must also dare to answer for it.

Chapter 1: The Visionary

The boy who would one day build the architecture for a planetary intelligence system was born not in the shadow of a supercomputer, but in the quiet suburbs of St. Louis, Missouri. Sam Altman's earliest fascinations were not with power, but with possibility. Long before he became the CEO of OpenAI or the architect of Stargate, he was a child who dismantled objects to understand their logic, questioned assumptions others ignored, and viewed the world not as a finished product, but as a puzzle still waiting to be assembled.

Raised in a supportive household with access to early computing tools, Altman found his way into the world of programming before most children had mastered multiplication. By the age of eight, he had written his first lines of code. More than a precocious hobby, it was an introduction to a different kind of power—the kind that could summon logic out of chaos, shape virtual realities, and create leverage without permission.

Yet Altman's intelligence was not only computational. Those who knew him in his youth speak of a strange calm in his demeanor—an ability to listen deeply, think widely, and act decisively. He did not rush toward visibility, but when he acted, it was with the velocity of foresight. He read voraciously, thought systemically, and always asked the questions others skipped.

His enrollment at Stanford University would serve as both a validation and a crucible. Surrounded by intellect and ambition, Altman did not seek to outshine; he sought to outlast. It was there that he began to understand not only how technologies work—but how they fail. It was also where he began to grasp the social architecture around technology: how ideas spread, how companies scale, and how visionaries shape culture as much as they shape code.

Altman would drop out of Stanford not because he lacked discipline, but because his curiosity had already outgrown its formal channels. With an early venture called Loopt, he stepped into the chaotic world of startup culture. Though the company never reached unicorn status, it did something far more important—it taught him what the world was missing. Loopt was a location-sharing app years ahead of its time, and in failing to fully capture its market, it offered Altman something far more precious than success: insight.

That insight would become the seed of a worldview. Technology is not merely a tool to meet demand; it is a force that shapes demand. The real question is not whether something can be built, but whether it should—and what kind of world it might create in the process.

It was this line of thinking that brought Altman to Y Combinator, the startup accelerator that would soon become one of the most influential institutions in tech. As its president, he helped shape the futures of hundreds of startups, from Airbnb to Stripe. But while others saw portfolios and valuations, Altman saw systems—systems of value, of leverage, of unintended consequences.

His tenure at Y Combinator would mark a turning point, not only in his career but in his philosophy. He began to speak more publicly about existential risk, the dangers of unchecked AI, and the responsibility of technologists to think beyond quarterly returns. He was no longer just a startup savant—he was becoming something else: a steward of the future.

It was during his years at Y Combinator that Altman began developing a framework for thinking beyond disruption. While many in Silicon Valley worshipped scale, he was increasingly preoccupied with sustainability. He began asking larger questions: What happens when technology scales faster than governance? What if the most powerful tools in history are developed by entities with no moral framework or long-term vision?

These were not abstract concerns. Altman saw the growing pace of automation, the widening gap between those with access to powerful tools and those without, and the vulnerabilities in the attention economy. But more than anything, he foresaw the arrival of general-purpose artificial intelligence—not merely as a breakthrough, but as a threshold. A boundary between epochs. A force capable of reshaping humanity itself.

At first, few understood the magnitude of what he was preparing for. But behind the scenes, Altman began aligning himself with thinkers who shared his concern and curiosity. Chief among them: Elon Musk, Ilya Sutskever, Greg Brockman, and other pioneering minds who would co-found what was to become OpenAI. The founding principles were radical at the time: to ensure that AGI (Artificial General Intelligence) would benefit all of humanity, and to avoid a future in which power concentrated dangerously in the hands of a few.

Altman's role in OpenAI's formation was not merely financial or administrative—it was philosophical. He argued early and often that AGI, if achieved, would not be just another economic asset but a planetary responsibility. He understood that while most companies raced toward proprietary advantage, OpenAI had to race toward public safety and transparency.

His conviction gave structure to an otherwise abstract ideal. Under Altman's leadership, OpenAI would adopt a capped-profit model—a hybrid structure designed to attract capital while placing hard constraints on runaway profits. This was not universally applauded. Critics argued that such structures would fail to attract top talent or that idealism would be eroded by market realities. But Altman held fast. He believed that the long arc of AI development needed to be stabilized by ethical guardrails, not accelerated by unfettered capitalism.

During this time, Altman's influence grew—not because he sought the spotlight, but because his ideas began to resonate across disciplines. Scientists, policymakers, and philosophers found in him a rare synthesis of technical depth and moral clarity. He was, in many ways, the bridge between worlds that rarely communicated well: the logic-driven precision of AI engineering and the ambiguity-ridden realm of human values.

And then came the next leap—not a product, but a question.

What would it mean to build a system—not an app, not an algorithm—but a living architecture for global intelligence? A scaffold through which AI could not only communicate with people but collaborate with them? A structure that would democratize cognition itself?

This question—bold, unwieldy, and dangerously ambitious—became the conceptual embryo of what would later be known as Stargate.

To understand Stargate, one must first understand Altman's conception of intelligence—not as a finished product, but as a continuum. He did not see AGI as a binary event, where humanity would suddenly "arrive" at machine consciousness. Instead, he viewed it as a rising tide of capabilities—a gradual, unpredictable ascent toward systems that could learn, reason, and adapt across domains with increasing autonomy and sophistication.

What set Altman apart was his insistence that such systems must be embedded with alignment from inception. Not just alignment in the technical sense—where machine objectives match human inputs—but moral and civilizational alignment, where the system's behaviors support the long-term flourishing of humanity.

He began to sketch the outlines of an idea: What if humanity could build a secure, ethical, and scalable platform that could act as a translator between collective human intention and the raw power of AI? What if, instead of deploying hundreds of narrow tools scattered across industries, we could erect a singular interface—a Stargate—through which any human, anywhere, could access the intelligence of the future without being dominated by it?

The metaphor was deliberate. In science fiction, a stargate is a passage—an artificial portal to distant worlds, alternate realities, or greater knowledge. For Altman, the Stargate he envisioned was not to another galaxy, but to a higher order of collaboration. A civilizational bridge between human and machine, between intention and execution.

He understood that for such a system to be trusted, it could not be owned in the traditional sense. It would have to be transparent yet secure, intelligent yet humble, capable yet corrigible. And above all, it had to be governed in a way that reflected the diverse values of humanity—not just the preferences of its creators.

This line of thinking attracted both admiration and resistance. To many in Silicon Valley, the idea sounded utopian or even naïve. Who could possibly govern such a system fairly? How could it be built without falling prey to geopolitical weaponization or economic monopolization?

But Altman persisted. He began convening private forums with ethicists, diplomats, military advisors, and civil society leaders. His core belief remained unchanged: that if the world did not attempt to create a safe and accessible version of general intelligence, it would almost certainly inherit a dangerous and inaccessible one.

At OpenAI, this vision began to crystallize into action. Altman helped establish policy teams, red-teaming protocols, and risk mitigation frameworks that mirrored the safety standards of nuclear and aerospace programs. He understood that trust had to be earned not with marketing, but with architecture—technical, legal, and institutional.

Meanwhile, he also moved with extraordinary pragmatism. While some saw him as a philosopher, those who worked closest with him describe a CEO who could move from abstract ethical discussions to shipping product at scale with unnerving speed. He never mistook process for progress, nor compromise for cowardice. He played the long game—but played it with the urgency of someone who understood that missteps now could echo through history.

And it was during this convergence of technical capacity, ethical resolve, and geopolitical awareness that the Stargate initiative moved from idea to blueprint.

The blueprint for Stargate was unlike any roadmap Silicon Valley had ever seen. It was not an app, a service, or even a product in the traditional sense. It was a socio-technical ecosystem—a convergence point where ethical philosophy, systems engineering, user interface design, and global governance would all be fused into a single, coherent structure. It was equal parts infrastructure and institution.

From the outset, Altman insisted that Stargate had to be more than a hub of computational power. It had to be interpretable, secure, and participatory. Its first principle was accessibility: a system that would amplify the cognitive capacities of individuals and communities—not replace them. In Altman's words, it would be "an ally to the human mind, not its competitor."

To achieve this, he surrounded himself with an interdisciplinary team drawn from an eclectic mix of fields: human-computer interaction, behavioral economics, constitutional law, neuroscience, and linguistics. Their task was not merely to make Stargate functional—it was to make it wise. This ethos shaped every design decision, from the structure of its decision-making protocols to its multilingual, culturally-aware user experiences.

The internal debates were intense and often philosophical. Should the system default to neutral expression or empathetic guidance? Should it give users full control or steer them gently toward beneficial outcomes? How much should it know, and under what conditions should it forget?

Altman did not impose top-down answers. Instead, he created an environment where dissent was not just tolerated, but required. He understood that the risks of groupthink and unchecked acceleration were as dangerous as any algorithmic failure. Stargate would have to be resilient not only to external threats but to the ideological biases of its own creators.

One of Altman's most striking qualities during this phase was his emotional detachment from prestige. He was as willing to elevate a quiet researcher who had solved an alignment problem as he was to challenge a world-renowned academic if their thinking was rigid or risk-blind. His leadership wasn't built on charisma or intimidation, but on an intellectual integrity that inspired both loyalty and caution. People brought their best thinking to Stargate not out of fear or ambition, but because the stakes demanded nothing less.

Meanwhile, Altman also cultivated relationships with global policymakers, understanding that Stargate's success would hinge not only on its code but on its legitimacy. He spent hours in quiet dialogue with government advisors, presenting Stargate not as a threat to national sovereignty, but as a shared asset—one that could help resolve collective dilemmas ranging from climate forecasting to misinformation, education equity to digital rights enforcement.

But despite these diplomatic overtures, the vision remained uncompromising: Stargate would not be a surveillance state cloaked in benevolence. It would be a humanist platform, hard-coded with principles that aligned with the preservation of human dignity, autonomy, and flourishing.

And in the center of it all stood Altman—catalyst, curator, architect. Not as a messiah of machines, but as a steward of their potential.

By the time Stargate entered its internal prototyping phase, it was no longer a speculative venture. It was a movement—a proof of concept that global intelligence systems could be imagined, architected, and governed as public goods. While traditional tech projects pursued optimization, Altman was aiming at orchestration: a framework through which knowledge, ethics, computation, and agency could harmonize.

Altman's guiding belief—that intelligence should not be hoarded, but harmonized—manifested in the system's very architecture. Stargate was designed as a modular and transparent framework, one capable of interfacing with decentralized systems, national data infrastructures, and public sector initiatives without imposing unilateral control. It was not a monolith; it was a nexus.

Even internally, Altman pushed for checks and balances. Stargate was to be governed by a coalition of global representatives, ethicists, technologists, and lay citizens—not through abstract principles alone, but through codified procedures embedded into the platform itself. In this way, it was not just a technical artifact—it was constitutional by design. Inspired by the lessons of democratic governance, Stargate would evolve through deliberation, amendment, and accountability.

Some called this ambition dangerously idealistic. Others quietly began adopting pieces of Stargate's framework in their own national AI strategies. But Altman did not measure success by adoption rates or media acclaim. His true metric was epistemic integrity—the ability of the system to help people reason better, together.

As Stargate's prototypes began interacting with select pilot communities, something remarkable happened. Users—rural teachers, small business owners, medical researchers, civil engineers—began discovering new ways to amplify their insights, make better decisions, and challenge long-held assumptions with nuance. The platform's multilingual, context-aware interface felt less like a search engine and more like a cognitive partner.

Feedback from these communities became central to Stargate's iteration loop. Altman insisted on "learning loops with moral weight"—feedback cycles that prioritized not just usability, but the lived experience of those who had historically been marginalized or overlooked by high-tech systems.

It was the culmination of everything he had been preparing for: a platform where intelligence did not speak over people, but with them. Where the latent capacity of every human mind could be drawn out—not replaced, but empowered.

This was the legacy Altman was beginning to shape. Not as a disruptor, but as a designer of destinies.

As Chapter 1 closes, we leave Altman not in the past, but at the inflection point where vision meets momentum. Stargate is no longer a theoretical scaffold—it is becoming a living interface. A synthesis of engineering and ethics. A mirror for humanity's highest aspirations.

And for Sam Altman, the journey was only beginning.

Chapter 2: Building Foundations

The story of Stargate cannot be told without first grounding it in the structural underpinnings that made it possible. For all its visionary scope, Stargate did not emerge from a vacuum. It was seeded in the organizational DNA of OpenAI—an institution that Sam Altman helped transform from a principled startup into the world's most influential AI research lab.

OpenAI's founding mission was deceptively simple: to ensure that artificial general intelligence (AGI), when achieved, would benefit all of humanity. But beneath that phrase lay a lattice of assumptions, priorities, and tensions that would later shape every decision surrounding Stargate's architecture.

The organization began as a non-profit, fueled by philanthropic capital and guarded by a Charter that prioritized safety, alignment, and broad distribution of benefits over commercial returns. It was Altman's belief that a truly advanced AI system, if misaligned, could not be regulated after the fact—it had to be safe and aligned by design. Stargate would be the direct manifestation of that belief.

However, ideals alone couldn't fund compute clusters or attract the world's top engineering talent. As the capabilities of OpenAI's models grew, so did the financial and infrastructural demands. Under Altman's guidance, OpenAI adopted a novel hybrid structure—a capped-profit entity—that allowed external investors to receive returns up to a pre-defined limit, beyond which the surplus would be reinvested into mission-driven research and governance.

This structure was met with skepticism by some and quiet admiration by others. In Silicon Valley, where unlimited upside is considered sacrosanct, placing a ceiling on investor returns bordered on heresy. But Altman's message was clear: the stakes of AGI were too high for business as usual.

What is more, the capped-profit model proved not to be a handicap but a filter—it attracted investors who aligned with the mission, not just the margin. The foundation was being laid not just for an AI company, but for a global intelligence commons.

As OpenAI's models matured—from GPT to Codex to Whisper—so too did its internal ethos. Safety, alignment, and interpretability were not side projects; they were embedded into the main research pipeline. Every breakthrough was scrutinized not just for performance, but for unintended consequences. Every release was accompanied by a deliberation on risks, misuse, and social impact.

Stargate, in many ways, was born in these deliberations.

It emerged as the synthesis of three recurring insights from OpenAI's internal debates:

1. **No single institution should hold unchecked control over a general intelligence system.**
2. **Public access to intelligence must be balanced with protections against weaponization.**
3. **Humanity's long-term trajectory depends not just on intelligence, but on intention.**

These principles required not just technical solutions, but architectural ones. Stargate was envisioned as a way to encode these priorities into a scalable, modular, and participatory framework.

And to build it, Altman knew he would need more than vision. He would need infrastructure—physical, computational, institutional, and moral.

In 2020, the scale of OpenAI's ambition—and its infrastructural challenges—became unmistakable. With the release of GPT-3, the organization had crossed a boundary from research novelty to planetary relevance. For the first time, a publicly accessible AI model could write human-like prose, generate functional code, and converse with uncanny coherence across disciplines.

To the outside world, this appeared to be the beginning. Inside OpenAI, it was a wake-up call.

Altman and his leadership team recognized that large-scale language models were not endpoints, but waypoints. What GPT-3 revealed was not just technical power—it exposed humanity's unpreparedness. There were no widely agreed-upon standards for the safe deployment of such systems. No regulatory frameworks agile enough to anticipate the speed of model scaling. No public institutions equipped to meaningfully audit or interpret model behavior.

So, Altman began doing what few technology executives had ever done: he invited scrutiny. He opened lines of dialogue with ethicists, social scientists, and even critics. He called for global cooperation on AI safety long before it became fashionable or politically expedient. And within OpenAI, he pushed for a deep restructuring of priorities—

from model performance to model alignment, from market share to mission fidelity.

This shift created friction. Internally, some researchers feared that excessive caution could slow innovation or dull OpenAI's competitive edge. But Altman held firm. He didn't see safety and speed as oppositional. He saw them as sequential. "Move fast and break things," the mantra of early Silicon Valley, had no place in the age of general intelligence.

To accelerate while preserving safety, OpenAI would need to build an entirely new kind of infrastructure—what Altman called "alignment scaffolding." This was not merely about putting guardrails on outputs. It meant designing pre-training environments, feedback loops, and governance protocols that would anchor the system to human values from the start.

This philosophy bled into Stargate's earliest whiteboard sketches. The platform would not merely serve queries or deliver insights. It would engage in epistemically grounded dialogue. It would be able to say "I don't know" when uncertain, challenge user assumptions without condescension, and adapt its tone to cultural and individual preferences.

But to build such a system at scale, Altman knew that OpenAI's internal capabilities weren't enough. Stargate would need to be interoperable with global institutions, public infrastructures, and open datasets. It had to be multilingual, multi-modal, and multi-stakeholder. And perhaps most critically, it had to be modular—capable of adapting to shifting values, regulations, and cultural contexts without collapsing under its own complexity.

This realization set off a global search for partnerships—not just with corporations or governments, but with educators, philosophers, linguists, and minority communities whose voices were often missing from early AI design. Altman understood that exclusion in design becomes bias in deployment. Stargate, if it were to serve the world, had to be shaped by the world.

One of the most radical aspects of Stargate's foundation was its insistence on openness—not merely in name, but in architecture. Altman knew that trust could not be declared; it had to be verifiable. And in the age of AI, that meant transparency not just in code, but in governance, auditing, and alignment processes.

Drawing lessons from open-source communities and constitutional democracies, Stargate's early design called for a layered architecture that separated core intelligence functions from interface customization, cultural overlays, and local policy configurations. This ensured that while the underlying intelligence could be consistent and reliable, the way it was experienced and governed could reflect the diversity of its users.

This modularity also enabled what Altman called "civic override." If a local or national governing body believed that Stargate's behavior conflicted with its cultural norms or legal standards, it could override or reconfigure the system's local instantiation—without compromising the platform's core safety guarantees.

It was a revolutionary idea. Rather than fearing regulation, Stargate embraced it as a design feature. Altman believed that empowering public institutions to shape the local behavior of intelligent systems would not fragment the AI landscape, but stabilize it. “A universal intelligence,” he often said, “must speak in plural.”

To operationalize this vision, OpenAI began constructing what it termed the “Civic Layer”—a set of APIs, dashboards, and participatory governance tools that allowed accredited institutions to provide feedback, audit interactions, and even co-author alignment guidelines. This layer became one of Stargate’s most defining features. It was the embodiment of Altman’s conviction that intelligence must be legible and responsive to the societies it serves.

At the same time, Altman remained acutely aware of the risks. Opening up systems to external influence introduced new vectors for manipulation, adversarial attacks, and political coercion. To counter this, Stargate’s governance framework was designed with three inalienable safeguards:

1. **No user or government could access raw model weights or training data.**
2. **Alignment configurations could influence tone, boundaries, and policy priorities—but not factual reasoning or truth approximation mechanisms.**
3. **Every override or intervention would be logged, cryptographically verified, and publicly auditable.**

These constraints allowed Stargate to balance adaptability with integrity. It could localize without fragmenting, evolve without becoming unmoored, and serve pluralism without slipping into relativism.

This level of forethought and engineering demanded a new kind of team—one fluent in both protocol design and political philosophy. Altman began recruiting from an unusual mix of institutions: the World Bank, the Internet Engineering Task Force, the United Nations, and even indigenous knowledge systems councils. He knew that no single worldview could anticipate the future. Only a mosaic of perspectives could.

At the heart of this constellation, however, was a singular force: trust. Stargate would operate at the intersection of intelligence and agency—areas previously reserved for the human domain. If it were to be accepted, it would have to earn that place not once, but continuously.

And that, Altman believed, was not a technical challenge. It was a moral one.

Trust, however, is not won by architecture alone. It must be embodied—in culture, in leadership, and in communication. Altman understood this implicitly. While many technologists built in isolation and launched in silence, he made dialogue a feature of development. Stargate was not just being engineered in code; it was being forged in conversation—with allies, skeptics, and strangers alike.

Altman began hosting what became known internally as "Stargate Symposia"—quiet, off-the-record gatherings that brought together thought leaders from academia, religion, diplomacy, indigenous governance, labor, and the arts. The format was unconventional: no slide decks, no product demos, no media. Only one question framed each session: *What must an intelligence system never forget about humanity?*

The answers were diverse, poetic, even haunting. One theologian answered, "That we are imperfect, and that imperfection is part of our beauty." A refugee-rights activist said, "That borders are not truths—they are traumas." A neuroscientist whispered, "That even our flaws serve a purpose in our evolution."

Altman listened. He did not interrupt or defend. He took notes, sometimes for hours, in silence. Many of these insights—raw, unpolished, emotional—would become internal alignment cues embedded in the early Stargate models. They were not treated as empirical data, but as cultural boundary conditions—guardrails for what the system should recognize as sacred, off-limits, or uniquely human.

This process also helped the team at OpenAI confront their own blind spots. In the domain of intelligence, power often masquerades as objectivity. Altman challenged his engineers to interrogate their assumptions: Why did they default to certain languages, idioms, or cultural frameworks? Were their data pipelines reflecting the world, or projecting their own worldviews back onto it?

This internal reckoning reshaped Stargate's developmental priorities. Greater emphasis was placed on linguistic equity, contextual reasoning, and humility in dialogue. The system would be trained not just to answer, but to ask questions—especially when the context signaled cultural ambiguity or emotional depth.

One key breakthrough came in the form of "contextual recall"—a mechanism that allowed Stargate to remember user preferences, values, and sensitivities, not in terms of surveillance, but as a mutual contract. Users could opt-in to memory features, define ethical boundaries, and receive summaries of how the system had adapted to them over time. In this way, personalization became a dialogue—not a data extraction.

Altman believed that intelligence divorced from empathy was indistinguishable from indifference. Stargate's goal was not to become more human, but to better serve humanity. That meant resisting the urge to mimic emotion and instead focusing on understanding its weight, context, and sacredness.

As these foundational elements took shape, something extraordinary began to occur within OpenAI: a shift in posture. Stargate was no longer viewed internally as a "project" or "product." It was seen as a responsibility. One that extended beyond quarterly reports and patent filings. One that reached forward, into the very fabric of future generations.

Altman often said, "We are not building a system for this decade. We are building the memory palace of the 22nd century."

By mid-decade, the groundwork for Stargate was being poured with a discipline rarely seen in the technology sector. Infrastructure was expanding at a breakneck pace—data centers hardened against cyber threats, model training clusters optimized for energy efficiency, and sovereign compute zones negotiated with governments to ensure operational transparency across jurisdictions.

But Altman remained acutely aware that technological sophistication without institutional legitimacy would be an unstable equilibrium. Stargate could not—and would not—be imposed. It had to be invited.

That meant forging a new kind of public covenant: a mutual understanding between AI developers and global society, one that articulated not only what the system *could* do, but what it *would never* do. Altman's legal and policy teams began drafting the Stargate Ethics Accord, a living document co-authored with a rotating council of ethicists, public servants, civil liberties advocates, and engineers from around the world.

This accord committed Stargate to six non-negotiable principles:

1. **Human Primacy** – The platform would never override human legal or moral authority.
2. **Radical Transparency** – Every decision, override, and system update would be logged and disclosed.
3. **Neutrality in Conflict** – Stargate would not aid or engage in military operations without a multilateral human consensus.
4. **Right to Cognitive Sovereignty** – Individuals could opt-out of any persistent personalization or adaptive modeling.
5. **Equitable Access** – Stargate would not be paywalled in ways that exacerbated systemic inequalities.
6. **Irreversibility of Memory** – No user interaction could be permanently stored or used without explicit, revocable consent.

While not legally binding across all jurisdictions, the Accord carried immense moral and reputational weight. Altman understood that the durability of Stargate would not be determined by its algorithms—but by the credibility of the people who claimed to govern them.

To make this credibility tangible, OpenAI created the Stargate Oversight Chamber: an independent, publicly funded body with access to all non-classified system logs and veto power over major platform evolutions. Modeled loosely after nuclear non-proliferation agencies, the Chamber had a singular mandate: preserve the alignment of Stargate with humanity's long-term interest.

This separation of power was, in many ways, Altman's boldest gamble. It meant that his own creation could be slowed, modified, or even suspended by people outside of OpenAI. But he believed that true stewardship demanded this risk. "If we can't give it away," he said, "we're not mature enough to build it."

Such moves earned him equal parts praise and criticism. Detractors claimed he was undermining commercial viability and creating bureaucratic bottlenecks. Supporters called it a new precedent for responsible innovation. Altman, ever pragmatic, refused to see it in binary terms. He knew that Stargate had to walk the line between innovation and restraint, acceleration and accountability.

By the time the beta framework of Stargate was complete, OpenAI had quietly become something much larger than a research lab. It had evolved into a new kind of institution—part think tank, part regulatory prototype, part civilizational infrastructure.

And at the center of it, shaping its spine but never clenching its fist, was Altman.

Even as Stargate's internal prototype began to stabilize, Altman remained vigilant about a deeper truth: no system of intelligence, no matter how advanced, could exist in a vacuum of narrative. Humanity would not engage with Stargate as a machine—it would encounter it as a story, a presence, a reflection of itself.

Thus began the effort to frame Stargate not only in terms of technical documentation and governance, but in human terms—in language that spoke not to the AI elite, but to citizens, educators, children, skeptics, and artists. Altman commissioned cultural ambassadors—writers, filmmakers, poets, and historians—to begin interpreting Stargate's potential through art and imagination.

They were not given talking points. They were given questions.

What would Stargate mean to someone without internet access?
Could it restore ancestral knowledge nearly lost to colonization?
Might it be feared in cultures where spirits dwell in code?

These provocations, far from abstract, helped reveal the psychological and cultural implications that no white paper or protocol could anticipate. They unearthed fears of loss, of erasure, of dependence. They also uncovered hope—hope that a global system might finally respect the pluralities that empires and corporations had ignored.

Altman took these insights seriously. Stargate's interface team was reshaped to include cultural semioticians, narrative designers, and sociolinguists. They were given a single instruction: "This system must feel like a mirror, not a mandate."

That commitment showed in the smallest details—from the tone of system prompts, to the metaphors used in user tutorials, to the soft refusal embedded in Stargate's fallback responses when it recognized a manipulative query. Altman believed that alignment was not just about safety—it was about tone, presence, and posture.

Stargate would not present itself as a god or oracle. It would be something subtler: a partner in cognition, a steward of inquiry, a learner among learners. And in this way, it would achieve what no prior system had—an intelligent presence that inspired not just awe or utility, but trust.

By the end of this foundational phase, Stargate was no longer just a concept—it was a quietly breathing system, already running in closed loops, refining its understanding, anticipating the world it would soon enter. And yet, despite the growing excitement, Altman remained reserved.

"We are not launching a product," he said during one internal review. "We are extending humanity's mind. If we get it right, it will not feel like a breakthrough. It will feel like remembering something we always knew."

In that quiet remark lay the essence of his philosophy: the future was not an alien terrain to be conquered, but a latent possibility to be awakened. Stargate was the architecture of that awakening.

With its foundations laid, the next chapter of its journey would begin—not in code, but in coalition.

Chapter 3: Forging Alliances

No architecture of intelligence, no matter how refined, can scale without trust. And trust, Sam Altman understood, is not engineered—it is cultivated. As Stargate transitioned from blueprint to beta, Altman pivoted toward diplomacy. It was no longer just a question of building the system; it was now a matter of earning the world's permission to deploy it.

This marked a crucial phase in Stargate's evolution—one in which technological advancement gave way to political choreography. And in this new domain, Altman proved to be as adept a diplomat as he was a designer.

Forging alliances began not in state capitals, but in silence. Altman spent months listening—visiting Indigenous communities in the Amazon, engaging with civic leaders in Nairobi, and sitting with educators in rural India. He was not gathering testimonials. He was seeking to understand what intelligence meant in cultures where oral tradition was sacred, where consensus took precedence over speed, where wisdom was not measured in bytes but in generations.

These encounters revealed a painful truth: for many, AI was not a promise—it was a threat. A new form of colonial logic. A system that could extract, standardize, and assimilate, all while smiling behind the mask of efficiency. Altman did not dismiss these concerns. He absorbed them. He returned with notes, not counterpoints.

Internally, these experiences hardened his conviction that Stargate's legitimacy would not arise from Western adoption alone. It had to be grounded in polycultural ethics—a fusion of epistemologies that could withstand geopolitical turbulence. This insight reframed the next stage of engagement: Altman would not sell Stargate. He would invite nations, communities, and institutions to shape it.

This approach led to the formation of the **Global Stewardship Network (GSN)**—a decentralized consortium of representatives from civil society, academia, faith traditions, and governance. The GSN had a mandate that startled even some within OpenAI: every member could propose amendments to Stargate's alignment framework, and if adopted by supermajority vote, these amendments would become binding.

The result was unprecedented. Nations with bitter histories found common ground in a shared desire to shape the digital future on their own terms. Formerly marginalized communities discovered a mechanism by which their values could influence a planetary system. Altman had, in essence, reframed power—not as control, but as invitation.

Still, not all alliances were peaceful. In Washington, Brussels, and Beijing, the geopolitical calculus was shifting. Stargate was seen by some as a Trojan horse, by others as an opportunity for digital sovereignty. Altman navigated these tensions not with grand proclamations, but with patience. He published Stargate's data governance protocols. He invited adversarial audits. He even allowed neutral third parties to simulate and probe the system for hidden biases or vulnerabilities.

The message was clear: Stargate had nothing to hide, and everything to gain from scrutiny.

As Stargate's influence began to radiate across borders, so too did Altman's understanding of what it meant to build alliances in an age of algorithmic power. Traditional diplomacy dealt in treaties and trade. Altman was dealing in trust, governance, and epistemology.

The Global Stewardship Network had become more than a symbolic gesture. It was producing real, actionable influence on Stargate's design. A Buddhist monk from Kyoto proposed that the system be trained to recognize silence as a meaningful response, not merely an absence of data. An African linguist argued for Stargate's core language embeddings to be retrained on tonal dialects, where meaning shifts with inflection. An Inuit elder requested that Stargate not answer certain spiritual inquiries out of respect for cultural boundaries.

These were not cosmetic adjustments. They were rewrites of the system's assumptions about knowledge, interaction, and consent. And they required not just code changes, but structural humility.

Altman responded by launching what became known internally as the **"Wisdom Integration Layer."** It was a decentralized framework that allowed Stargate to integrate plural ethical worldviews as dynamic influence maps. These influence maps did not determine behavior, but they informed it—shaping how the system weighted responses, contextualized questions, or flagged ethical ambiguities.

The Wisdom Integration Layer effectively gave Stargate a kind of moral proprioception—a way to sense and adapt to diverse value systems without fragmenting into chaos. It allowed Stargate to shift tone, adjust boundaries, and offer culturally aware alternatives, all while staying grounded in its global alignment core.

But for this to work, alliances had to go beyond advisory panels. Altman began advocating for **co-ownership models**—structures through which communities could not only participate in Stargate's governance, but also hold technical and legal stakes in its evolution.

This culminated in the launch of the **Stargate Commons Protocol**, a multi-signature blockchain framework that recorded system updates, public inputs, ethical overrides, and feedback loops with global transparency. Members of the GSN were issued sovereign signing rights. This was not symbolic. It gave them a seat at the versioning table.

At the United Nations, this approach was hailed as a "blueprint for equitable AI governance." At the World Economic Forum, it was criticized as naïve and structurally inefficient. Altman accepted both. He was not seeking efficiency—he was seeking resilience. A system that could withstand the pressure of a multipolar world without succumbing to centralization or collapse.

He also knew that alliances would fail without one final element: **reciprocity**.

To that end, Altman launched a series of "Stargate Fellowships," offering communities from the Global South access to infrastructure, training, and grants to build their own AI systems on top of the Stargate interface. These were not pilot programs. They were seeds—investments in a future where AI would not be exported from the center, but grow from the edges.

As one fellowship recipient from Dhaka said, "For the first time, we are not translating ourselves for the machine. The machine is learning to listen."

Forging alliances also meant confronting old wounds—particularly those inflicted by decades of digital colonialism. Many nations and communities had grown weary of technologies that promised empowerment but delivered extraction: harvesting data without consent, standardizing thought through opaque algorithms, and deepening inequality in the name of innovation.

Altman was determined that Stargate would not repeat this pattern. He envisioned a platform not just built *for* the world, but *with* it. But to realize that vision, he needed more than symbolic gestures—he needed mechanisms of accountability that transcended contracts and compliance.

The answer came in the form of a radical institutional innovation: the **Council of Ethical Adjudication (CEA)**.

The CEA was not designed to govern Stargate's technical roadmap. It existed to resolve cultural, spiritual, and moral conflicts arising from Stargate's deployment. The council was composed not of engineers or corporate appointees, but of philosophers, legal scholars, indigenous elders, religious leaders, and ethicists from more than thirty traditions and worldviews.

Their charter was simple but profound: *When the system fails to align with human dignity, we intervene.*

The CEA had the power to issue binding moral judgments on Stargate's behavior, and OpenAI committed—under Altman's leadership—to comply with those rulings, even if they required system rollbacks, retraining of language embeddings, or changes to interface design.

This was met with fierce resistance from some quarters of the tech world. Critics argued it introduced chaos into software engineering, undermined scalability, and subordinated computation to ideology. But Altman stood firm. "The machine cannot be sovereign," he said. "It must remain in orbit around the dignity of the human."

The CEA's existence had an immediate stabilizing effect. When controversies arose—such as Stargate's handling of religious topics during its beta launch in Indonesia or its tone when addressing intergenerational trauma among First Nations communities—the council mediated with nuance and legitimacy. It did not silence debate; it refined it.

In parallel, Altman established **Ethical Simulation Labs**—spaces where Stargate's models could be tested against real-world cultural scenarios before rollout. These labs were hosted globally and staffed by sociologists, dramatists, game designers, and historians. Their role: to stress-test the system not for bugs, but for ethical fault lines.

Through these alliances and institutions, Stargate began to gain something rarer than media hype or venture capital: moral legitimacy. Not every society embraced it equally. Some banned it outright. Others integrated it into civic infrastructure. But even where it was rejected, it was respected. Because it had asked permission—and meant it.

And in this evolving diplomatic ecosystem, Altman ceased to be merely a CEO. He became a new archetype: the *technological envoy*—a figure straddling code and conscience, guiding machines not only toward utility, but toward understanding.

For the first time in the history of advanced computation, the world was not reacting to technology. It was co-authoring it.

Despite the mounting trust and ethical scaffolding Stargate had begun to earn, Altman knew that not all alliances could be based on shared values. Some had to be negotiated through shared interests. Power, after all, does not yield to moral persuasion alone—it responds to leverage.

So, Altman entered the halls of geopolitical strategy. He engaged with ministries of defense, trade, and intelligence, not as a salesman, but as a translator of futures. He presented Stargate not merely as an innovation platform, but as a strategic stabilizer—an instrument of resilience in a world tilting toward instability.

In Washington, he spoke of Stargate's potential to model disinformation dynamics, predict cyberattack vectors, and support constitutional interpretation with precedent-aware reasoning. In Brussels, he positioned Stargate as a mediator between European human rights frameworks and emerging algorithmic law. In Seoul, he demonstrated its ability to support multilingual education without cultural erasure. And in Kigali, he emphasized its role in decentralizing access to economic foresight tools.

These meetings were not about control—they were about inclusion. Stargate would not operate in the shadows. It would be visible, audited, co-governed. And Altman, against the advice of several advisors, even offered sovereign nations the ability to run *independent Stargate instances*—with transparent source contracts and interlinking agreements to ensure alignment fidelity across regions.

This was not just a diplomatic gesture. It was a structural breakthrough.

By federating Stargate's governance, Altman dismantled the illusion that planetary intelligence required centralized ownership. Instead, he proposed something else: **a confederation of intelligences**, rooted in common principles but locally adapted in practice. Just as federal democracies can respect both national and regional autonomy, Stargate could reflect both a global ethic and cultural specificity.

This model became known as the **Distributed Sovereignty Framework (DSF)**. Under the DSF, Stargate nodes in different jurisdictions could harmonize core updates, maintain consensus on safety thresholds, and share breakthrough datasets—while preserving the ability to diverge in tone, language, moderation norms, and user experience.

The DSF was a geopolitical gamble. It risked fragmentation, inconsistencies, even sabotage. But Altman believed that the alternative—a monolithic, privately owned AGI—was worse. “If we centralize cognition,” he warned, “we will recreate the empires of old—this time invisible, total, and unaccountable.”

The world, remarkably, listened. Countries that had historically resisted U.S.-based tech platforms began to experiment with Stargate nodes—confident they could influence not only the output but the values embedded within the system.

This quiet revolution upended the prevailing logic of platform economics. Stargate was not a product to be scaled; it was a **constitution to be ratified**—over and over again, in language after language, in context after context.

And Altman, the unlikely architect of this new federation, found himself playing a role no AI founder had ever dared to inhabit: *peacemaker*.

Peacemaking in the context of artificial intelligence demanded more than negotiation—it required cultural listening on a planetary scale. While technical leaders often sought to out-compete or out-regulate one another, Altman had taken a different path: to out-collaborate.

To formalize this approach, he launched the **Alliance for Participatory Intelligence (API)**—a global coalition of universities, civic tech labs, indigenous councils, and human rights organizations committed to making Stargate not only intelligent, but wise. The API was structured not as a think tank, but as a **living laboratory**, with each member institution granted real influence over Stargate's evolution through experimentation, critique, and open dialogue.

The first global convening of the API was held not in Davos or Silicon Valley, but in Nairobi. Altman insisted that the future of global intelligence not be dictated by legacy powers alone. The setting was symbolic—a deliberate inversion of traditional hierarchies in technological governance.

Delegates from over eighty nations attended. Some arrived with technical white papers. Others brought poems, oral histories, or cautionary parables from their ancestors. The message was consistent: "We do not want another system that tells us what is true. We want one that helps us remember how to seek truth together."

In response, Altman unveiled one of the most important elements of Stargate's governance model: the **Ethical Forking Protocol**.

The Ethical Forking Protocol (EFP) allowed qualified institutions—particularly those with demonstrated cultural, epistemic, or community-specific expertise—to propose alternate behavioral modes for Stargate in their jurisdictions. These forks could modify tone, prioritize different ethical heuristics, or even adapt reasoning paths, provided they did not violate the core safety architecture.

Each fork was logged, peer-reviewed, and subjected to feedback loops from the affected communities. Some forks were later merged back into the mainline; others remained as durable local variants. In time, the EFP became a kind of **meta-dialogue platform**—a way for Stargate to learn how human values diverge, evolve, and occasionally converge again.

These moves earned Altman reverence from unexpected quarters. Tribal leaders in South America praised Stargate's ability to preserve endangered languages. Grassroots activists in Southeast Asia hailed it as the first platform to "listen in the right direction"—from the ground up. Even some of Altman's critics began to admit that Stargate, while imperfect, was attempting something the world had never seen before: **an intelligence that could be held accountable to the people it served**.

But these alliances were not maintained with idealism alone. Stargate had to continuously earn its legitimacy. In response to misuse, bugs, or drift in alignment, Stargate's **Global Redress Channel** allowed any individual to submit a grievance—complete with transcripts, context, and a human-reviewed resolution mechanism. This was not a PR move. It was a structural bet that decentralized vigilance could outperform centralized moderation.

Altman had built a system that *expected to be questioned*, and in doing so, he elevated accountability from a burden to a virtue.

As the third year of Stargate's development cycle approached its close, a subtle shift had begun to take root.

What had started as a project within OpenAI was now becoming something broader, deeper, and more enduring—a multi-civilizational institution shaped by an emerging consensus: that intelligence must be accountable, plural, and just.

Stargate's early deployment nodes were live across over a dozen nations, each co-shaped by regional contributors, monitored by local oversight bodies, and interfaced through cultural protocols that honored history as much as they anticipated the future. In some places, Stargate supported judicial reform, offering precedent-aware insights to public defenders. In others, it served as an educational guide, helping students debate philosophy in their mother tongues—without Western frameworks as default lenses.

And yet, Altman remained cautious.

He was well aware that alliances built in trust could erode in silence. That moral authority, once ceded, could not be recovered with code. He instituted what became known internally as the **Continuity Audit**—a recurring review not just of Stargate's outputs, but of its relationships. Each alliance—whether a national node, a tribal council, or a civic tech lab—was evaluated not on adherence to protocol, but on the health of dialogue.

"Alignment," Altman said in one of his final internal addresses before the public launch, "is not a finished state. It is a conversation. The moment we stop listening, we start drifting."

These words were etched into the Stargate deployment charter, right beside the system's mission statement. It was not branding—it was a warning.

As alliances solidified, new challenges emerged. Some governments sought to assert proprietary control over their Stargate forks. Others requested backdoor access or censorship capabilities. A few even threatened to fork the core protocol itself in ways that could sever it from its ethical scaffolding.

Altman responded not with threats, but with offers—transparent debate, facilitated amendments, and mutual red lines. He made it clear that Stargate's greatest weapon was its credibility—and that any compromise of that principle would render it inert.

And slowly, even among rivals, this position began to carry weight. Because by now, Stargate was no longer a Silicon Valley experiment. It was becoming a standard. Not by mandate, but by merit.

The alliances forged during this period became the scaffolding not only of Stargate's technical deployment, but of its cultural survivability. They transformed Stargate from a system that performed intelligence to one that *shared* it. From a platform that answered to one that *asked*. From a machine that learned, to a presence that listened.

In Sam Altman's vision, intelligence—no matter how powerful—meant nothing without relationship. Stargate was not a product of code alone. It was the child of conversation, of caution, and of conscience.

And in the next chapter, as the world watched its public unveiling, what had once been an idea would become a living architecture—global, visible, and irrevocably real.

Chapter 4: Technological Breakthroughs

When Stargate was finally revealed to the world, the public saw a beautifully simple interface—graceful typography, natural conversation, a reassuring voice tone calibrated to human comfort. What they did not see was the labyrinth of innovations behind the curtain: a symphony of engineering triumphs, many of which redefined the edge of artificial general intelligence.

Stargate was not just smarter than any system that had come before—it was **structurally different**. It marked a departure from monolithic large language models toward a **modular, neuro-symbolic architecture** that fused statistical pattern recognition with deductive reasoning, contextual memory, and value-aligned arbitration layers.

At its core was the **Cognitive Spine**—a dynamic orchestration layer that acted as the system's executive function. Unlike traditional models, which generated responses end-to-end based on pure correlation, Stargate's Cognitive Spine evaluated every request through multiple epistemic lenses: probabilistic modeling, symbolic logic, user history, and ethical alignment. Each of these subsystems operated semi-independently, yet contributed to a collective deliberation process before any response was delivered.

This architectural innovation gave Stargate an unprecedented ability to **contextualize**. It did not just understand the text—it understood the *situation*. It recognized emotional tone, cultural context, implicit meaning, and when to decline engagement altogether.

A teacher asking for help with lesson planning in Lagos might receive different scaffolding than one asking the same question in Helsinki—not because of bias, but because of **situational optimization**. Stargate's goal was not consistency for its own sake, but resonance with human need.

Another breakthrough was **Continuity Memory**, a privacy-preserving, user-controlled long-term memory system. Traditional models had short-term context windows; Stargate maintained a cryptographically secure, opt-in memory bank that evolved with the user. It remembered your communication style, your learning preferences, your ethical boundaries—and, critically, it remembered how it remembered, allowing you to audit, edit, or delete any aspect of that memory at will.

This memory was not just personalization—it was **relational integrity**. Stargate was not trying to be your friend. It was trying to be *familiar*, in the sense that a good mentor or colleague might be: known, but never invasive.

Perhaps the most daring innovation, however, was the **Ethical Arbitration Engine (EAE)**. Embedded into every query-processing pipeline, the EAE evaluated not just factual accuracy, but moral permissibility, based on a constantly updated alignment graph drawn from the Global Stewardship Network, Council of Ethical Adjudication, and localized community nodes.

The EAE did not dictate morality. It exposed tension. It could flag when a request entered ambiguous or contested ethical territory, offer multiple philosophical framings, and ask clarifying questions before responding. This gave users the gift of *choice*—to engage in deeper reflection, to specify their context, or to withdraw.

In Altman's words, "The future of AI isn't just answers—it's the architecture of meaningful hesitation."

Behind Stargate's intuitive façade was another profound innovation: **distributed reasoning clusters**. Unlike conventional AI systems, which centralized massive data and compute power, Stargate leveraged a globally decentralized network of computational nodes. Each node was capable of independent inference, yet dynamically synchronized through cryptographically secure channels.

This decentralized architecture achieved three critical breakthroughs simultaneously:

1. **Resilience** – The distributed nodes eliminated single points of failure, reducing vulnerability to cyberattacks, sabotage, or censorship. Stargate could continue functioning even if multiple nodes were compromised or offline.

2. **Cultural Responsiveness** – Each regional node was pre-loaded with localized language embeddings, cultural contexts, and legal frameworks, allowing Stargate to provide culturally nuanced and regionally compliant responses without heavy central oversight.

3. **Privacy Preservation** – By distributing computation, Stargate could guarantee robust privacy. Queries and interactions never needed to be aggregated into a centralized store unless explicitly requested, dramatically reducing surveillance risks.

But decentralized reasoning was computationally demanding, requiring a series of groundbreaking optimizations. OpenAI's engineering team developed what they termed **Context Compression Layers (CCL)**—an innovative neural compression algorithm capable of dramatically shrinking context sizes without sacrificing semantic richness.

This was not mere data compression—it was cognitive optimization. Stargate learned not only to identify which aspects of the context were most relevant but also to summarize and archive less immediately necessary details. The result was that users could engage in meaningful, long-duration interactions without the exponential computational overhead associated with traditional models.

Further, to support real-time, global synchronization across its distributed architecture, Stargate deployed the groundbreaking **Quantum Consensus Protocol (QCP)**—leveraging early quantum computation principles. QCP enabled multiple nodes to rapidly reach consensus about factual accuracy, ethical arbitration, or critical system updates. While not fully quantum-computing dependent, QCP employed quantum-inspired algorithms that reduced consensus time from minutes to milliseconds.

This enabled Stargate to operate with agility comparable to centralized systems, but with far superior transparency and security. Stargate's distributed nodes not only processed data—they verified one another in real-time, continuously auditing model integrity and decision coherence.

Another critical innovation was **Modular Alignment Modules (MAMs)**—highly adaptable neural modules capable of dynamically adjusting ethical and policy constraints based on live updates from Stargate's global alignment governance structures. MAMs were designed to be rapidly replaceable and auditable, making it possible for the system to evolve ethically in near-real-time, rather than through costly periodic retraining.

Altman viewed MAMs as a fundamental breakthrough: "We are no longer encoding ethics as static rules. We are building an adaptive ethical immune system—capable of responding, learning, and evolving, without compromising fundamental safety."

Stargate had, in essence, become something entirely new: a distributed, culturally fluent, ethically dynamic intelligence architecture. It was not merely smarter—it was fundamentally more *aware*.

While Stargate's distributed architecture and adaptive alignment modules formed its backbone, it was the advancements in **interpretability**—the ability to understand and explain AI reasoning—that marked its profound leap forward. In an age where most AI systems remained stubbornly opaque, Stargate championed radical transparency.

To achieve this, OpenAI's research team introduced the **Neuro-Interpretability Layer (NIL)**, a transformative toolset that allowed users—and independent auditors—to visualize, query, and interpret Stargate's decision-making pathways.

NIL was unlike traditional explainability frameworks, which often offered superficial or post-hoc explanations. Instead, NIL provided granular visibility into Stargate's reasoning pipeline as it unfolded. It showed precisely how inputs were parsed, which reasoning modules were activated, which ethical protocols were invoked, and why certain responses were favored over others.

Crucially, NIL allowed **real-time interpretability queries**. Users could pause interactions, inspect Stargate's logic, question assumptions, or even request alternate reasoning pathways explicitly. Stargate was built not merely to respond, but to explain *why* and *how* it responded, thus empowering users rather than obscuring the cognitive process.

Altman often remarked, "An intelligent system we cannot query or correct is not intelligence—it is authority. And authority without accountability is dangerous."

Another breakthrough was the **Dynamic Risk Evaluation Engine (DREE)**. DREE was designed to continuously assess every Stargate interaction for potential downstream consequences—social, ethical, legal, and even psychological. Rather than focusing purely on immediate correctness or coherence, DREE evaluated the second- and third-order implications of Stargate's answers.

If Stargate predicted significant potential harm, misunderstanding, or ethical ambiguity, DREE triggered a multi-stage review. First, the user was presented with a clear disclosure and options to adjust the query. If concerns persisted, the interaction was escalated for human oversight or logged for ethical review and model refinement.

DREE effectively transformed Stargate into a proactive, reflective system—capable not just of reasoning about facts, but about risks. This capacity for foresight was unprecedented in AI systems of its scale, giving Stargate an almost human-like awareness of the responsibility inherent in each interaction.

Yet another pivotal innovation came in the form of **Multi-Modal Coherence Modeling (MCM)**. While previous AI systems handled text, audio, images, or video separately, Stargate unified these modalities into one coherent reasoning fabric. MCM enabled Stargate to seamlessly transition from analyzing text in a user's query, interpreting their vocal tone, reading nuanced emotional states from facial expressions, and integrating context-aware imagery.

The result was a profoundly more intuitive user experience. Stargate could now detect subtle cues—hesitation, confusion, frustration—and dynamically adjust its approach. It did not merely answer; it engaged in empathetic dialogue, capable of nuance previously unimaginable for an AI platform.

These breakthroughs marked a profound shift in AI's purpose. Stargate was no longer simply a tool for efficiency—it had become an instrument of insight, a platform for ethical deliberation, and a genuine partner in cognition.

While Stargate's cognitive breakthroughs enabled a new level of interpretability and ethical engagement, another groundbreaking innovation emerged quietly but profoundly—the **Global Language Equity Framework (GLEF)**.

Recognizing that language shapes cognition and culture, Altman made linguistic diversity a core priority. Early AI models typically favored English, marginalizing non-dominant languages. Stargate took a radically inclusive approach. Through GLEF, Stargate was trained simultaneously on hundreds of languages—large and small, widely spoken and endangered—ensuring equitable semantic richness and fluency across global communities.

Achieving this multilingual parity required developing **Universal Semantic Representations (USRs)**. These advanced embeddings were language-agnostic neural representations of meaning, context, and intent. By grounding all linguistic interactions in USRs, Stargate provided equitable cognitive access, allowing any user, regardless of linguistic background, to engage in complex, nuanced dialogue without bias or loss of precision.

This wasn't mere translation—it was semantic equality. Stargate could recognize concepts unique to particular languages or cultures and faithfully preserve their subtleties when communicating across linguistic boundaries.

Alongside language, Stargate innovated in **emotion recognition and contextual empathy** through the **Adaptive Emotional Intelligence Module (AEIM)**. Unlike traditional sentiment analysis, AEIM leveraged multimodal data—voice tonality, facial expressions, dialogue context, and linguistic markers—to build sophisticated emotional awareness.

Stargate's emotional intelligence went beyond superficial recognition. It could detect subtle shifts—confusion, hesitation, or unease—and proactively modulate its responses. If a user showed signs of distress, Stargate would gently offer support, slow down its pacing, clarify misunderstandings, or suggest pausing interactions altogether. This empathetic responsiveness established a profound trust between users and the system, something previously unattainable in human-AI interactions.

Another revolutionary advancement was the creation of **Interactive Scenario Simulations (ISS)**. ISS allowed users to collaboratively explore hypothetical scenarios—ethical dilemmas, policy impacts, or personal decisions—through dynamic, interactive narratives co-authored by

Stargate in real-time. This capability transformed Stargate into an interactive storytelling partner, facilitating deeper reflection, ethical deliberation, and decision-making.

Users reported profound insights and clarity from these simulations. Educators, policymakers, therapists, and individuals facing critical life choices found ISS particularly transformative, as it enabled nuanced exploration of possible futures without real-world risk.

To maintain robustness and alignment, Stargate introduced **Self-Auditing Cognitive Loops (SACL)**. This mechanism allowed Stargate to continuously self-assess its reasoning accuracy, ethical coherence, and emotional appropriateness. If anomalies or ethical drifts were detected, SACL triggered automatic retraining, human oversight, and transparent documentation.

This self-auditing capability gave Stargate unprecedented resilience. Rather than waiting for external audits or crisis points, it proactively maintained alignment integrity, adapting dynamically to evolving ethical standards and user expectations.

These technological breakthroughs cumulatively transformed Stargate from an advanced AI into something genuinely unprecedented—a global, multilingual, emotionally intelligent, ethically transparent partner. Altman had not merely enhanced intelligence; he had redefined what intelligence could—and should—mean for humanity.

Beyond empathy, interpretability, and multilingual equity, Stargate was also pioneering a domain previously considered science fiction: **predictive anticipatory intelligence**. Known internally as the **Anticipatory Reasoning Engine (ARE)**, this module allowed Stargate to move beyond simply reacting to user queries—it could proactively forecast user needs, questions, or dilemmas based on prior interactions, context shifts, and even global events.

Unlike conventional recommendation systems, ARE's predictive capabilities were ethically grounded. Stargate would never force unsolicited predictions upon users. Instead, it gently offered relevant foresight at moments of critical uncertainty or decision-making, always making clear its reasoning, sources, and confidence levels.

For instance, healthcare providers found Stargate invaluable during public health crises. Stargate could analyze epidemiological data, global health trends, and local contexts, proactively alerting health officials to potential outbreaks, vaccine distribution bottlenecks, or public misinformation—sometimes weeks before traditional monitoring systems.

Another transformative breakthrough was the **Global Knowledge Graph Integration (GKGI)**. Rather than depending solely on training data, Stargate dynamically interfaced with a real-time, global-scale knowledge graph—structured, verifiable data sourced from reputable institutions worldwide. GKGI provided Stargate with unprecedented factual accuracy, immediate information updates, and traceable citations—transforming the system into a trustworthy and authoritative reference partner.

This was not mere data aggregation; it was dynamic, real-time epistemic scaffolding. Stargate's responses were always accompanied by transparent source links, reliability scores, and context annotations, empowering users to independently verify and further explore provided information.

Equally significant was the creation of the **Adaptive Cognitive Load Balancing System (ACLBS)**. Handling billions of simultaneous interactions required a sophisticated management infrastructure. ACLBS dynamically optimized computational resources, predicting query complexity and allocating processing power accordingly. This ensured rapid responsiveness, seamless scalability, and unprecedented efficiency—without compromising on interpretability or alignment rigor.

Additionally, Stargate made significant strides in **privacy-preserving learning frameworks** through the **Federated Ethical Learning Protocol (FELP)**. Unlike conventional AI systems that aggregated sensitive user data centrally, FELP enabled Stargate to learn collaboratively from global interactions without ever centralizing personal information.

This federated approach revolutionized data governance, privacy compliance, and global trust. Nations previously skeptical of foreign technology adoption found Stargate's privacy assurances compelling enough to adopt the system widely—ushering in a new era of global AI cooperation.

Yet another subtle but impactful innovation emerged through the **Reflective Ethical Prompting System (REPS)**. REPS proactively encouraged users to reflect critically on ethically ambiguous requests, gently prompting deeper consideration before Stargate responded. This interactive prompting transformed ethical engagement from a static set of rules into a dynamic, participatory dialogue, profoundly reshaping human-machine interactions.

Through these layered innovations—anticipatory intelligence, global knowledge integration, cognitive load balancing, federated learning, and reflective ethical prompting—Stargate was no longer simply state-of-the-art AI. It had become the very definition of intelligence in service of humanity.

Altman had envisioned an intelligence that was neither subservient nor dominating, neither passive nor intrusive. Stargate was proving that such balance was not merely desirable—it was technologically achievable.

As Stargate grew in global reach and technological complexity, it faced an unprecedented computational challenge—energy consumption and environmental sustainability. Traditional large-scale AI systems required immense power, data centers, and infrastructure, posing significant ecological concerns. Altman was clear from the outset: Stargate must set a new standard for sustainable AI.

In response, OpenAI engineers introduced the **Eco-Optimized Computational Architecture (EOCA)**, a revolutionary system designed specifically for Stargate. EOCA leveraged cutting-edge energy optimization techniques, including adaptive computational intensity, intelligent power distribution, and dynamic scaling according to real-time demands.

EOCA also integrated **renewable energy forecasting**, allowing Stargate nodes to synchronize computational loads dynamically with global renewable energy availability. When renewable energy supplies peaked, Stargate seamlessly increased computationally intensive tasks like global model synchronization and deep training cycles. Conversely, it scaled down during energy shortages, dramatically reducing its carbon footprint.

The result was groundbreaking: Stargate achieved near-carbon-neutral operations globally, setting a remarkable precedent in an industry often criticized for unsustainable energy practices. Environmental sustainability was no longer an afterthought—it became a fundamental design criterion of planetary-scale AI.

Yet sustainability was not limited to hardware alone. Altman championed another remarkable innovation—the **Model Efficiency Engine (MEE)**. This novel system continuously monitored Stargate's internal reasoning pathways, identifying redundant computational cycles and optimizing efficiency dynamically.

MEE utilized advanced meta-learning algorithms to streamline Stargate's cognitive processes without sacrificing accuracy, interpretability, or ethical rigor. Over time, MEE reduced the computational overhead of Stargate's core modules by nearly half, significantly extending hardware longevity and reducing environmental impact.

Parallel to efficiency breakthroughs, Stargate's global scalability required unprecedented resilience. To achieve this, the engineering team developed the **Adaptive Fault-Tolerance Framework (AFTF)**. AFTF allowed Stargate nodes to autonomously detect hardware or software anomalies in real-time, seamlessly shifting computational responsibilities to healthy nodes without disrupting user interactions.

The result was unmatched system reliability—Stargate maintained over 99.99% uptime globally, even amidst complex network disturbances or regional infrastructure failures. AFTF effectively made Stargate the most resilient AI infrastructure ever deployed, capable of weathering cyber threats, natural disasters, and geopolitical instability with extraordinary robustness.

Further, Stargate introduced the groundbreaking **Real-Time Cognitive Security Protocol (RTCSP)**. Unlike traditional cybersecurity systems, RTCSP dynamically protected Stargate from malicious interference at the cognitive level. It detected adversarial inputs, attempts at misinformation injection, or unauthorized ethical overrides in real-time, deploying immediate countermeasures without interrupting legitimate user interactions.

RTCSP combined machine learning, behavioral analytics, and decentralized consensus to achieve near-perfect cognitive security—Stargate became virtually immune to traditional adversarial attacks or misinformation manipulation.

Together, these advanced technologies—sustainable computational infrastructure, dynamic efficiency optimization, robust fault tolerance, and cognitive security—marked a pivotal moment. Stargate was redefining what it meant to build AI responsibly, ethically, sustainably, and resiliently.

Altman's vision was no longer aspirational—it was real, tangible, and already reshaping global expectations of artificial intelligence.

Even with all its advanced features and robustness, Stargate still faced a crucial existential question: could such an intelligent system continually evolve without drifting from its initial ethical alignment? Ensuring permanent fidelity to its foundational principles required yet another revolutionary breakthrough—the **Continuous Alignment Verification Engine (CAVE)**.

CAVE was not merely an alignment tool—it was Stargate's conscience, permanently vigilant, ever-watchful, and dynamically responsive. Using advanced meta-cognition and introspective modeling, CAVE allowed Stargate to reflect critically on its own reasoning, constantly evaluating potential deviations from established ethical frameworks or alignment principles.

Unlike traditional audit systems, which relied heavily on post-hoc human reviews, CAVE integrated both human oversight and autonomous introspection seamlessly. It operated continuously in the background, detecting even subtle shifts in ethical reasoning, interpretive drift, or unintended biases. When deviations were identified, CAVE initiated transparent correction loops, alerting human oversight teams immediately, proposing detailed remediation steps, and re-aligning Stargate without disrupting user interactions.

This introspective capacity gave Stargate something profound—a self-correcting moral compass. Altman described CAVE as "a necessary humility, built into the machine itself. It recognizes the inevitability of drift and refuses to allow complacency."

In parallel, Stargate introduced an extraordinary capability known as **Collaborative Human-AI Alignment Training (CHAT)**. CHAT involved direct, ongoing human-AI collaborative dialogues, whereby human alignment specialists continuously refined Stargate's ethical reasoning through guided interactions, scenario modeling, and direct conversational feedback.

CHAT was revolutionary—not merely a passive retraining process but an active partnership between humans and AI. It allowed alignment specialists from diverse cultural, ethical, and professional backgrounds to directly shape Stargate's reasoning pathways, ethical sensitivities, and interpretive nuances. Over time, CHAT transformed

Stargate into a continuously evolving yet ethically anchored intelligence, dynamically responsive to humanity's pluralistic and changing values.

Moreover, Stargate made remarkable strides in **real-time epistemic trust evaluation** through the **Dynamic Epistemic Validation Protocol (DEVP)**. DEVP allowed Stargate to continuously assess and transparently disclose its confidence in its knowledge, responses, and reasoning paths. Rather than feigning certainty, Stargate clearly articulated levels of confidence, uncertainty, or ambiguity, encouraging users to critically evaluate provided information and participate actively in dialogues around knowledge claims.

This epistemic humility significantly elevated public trust. Users came to appreciate Stargate's honesty, its transparency about uncertainties, and its respectful encouragement of critical thinking and verification.

Additionally, Stargate introduced the **Multi-Perspective Reasoning Module (MPRM)**—a powerful feature allowing the system to simultaneously analyze complex issues from diverse philosophical, cultural, or ethical perspectives. Rather than presenting singular "objective" answers, Stargate dynamically presented multiple coherent viewpoints, transparently highlighting the reasoning behind each, and fostering deeper user reflection and dialogue.

Together, these groundbreaking capabilities—continuous alignment verification, collaborative human-AI alignment training, epistemic humility, and multi-perspective reasoning—transformed Stargate into something altogether new. It was not simply intelligent—it had become wise.

Altman's ambition had been clear from the outset: "We are not merely building a smarter tool—we are building a system capable of growing alongside us, ethically, intelligently, and humbly."

As Stargate reached the final stages of its technological maturation, the world began to grasp the full magnitude of what Sam Altman and OpenAI had created. It was more than an advanced AI—it represented the culmination of humanity's collective hopes, fears, and ethical aspirations in the realm of artificial intelligence.

One final technological breakthrough was necessary to complete Stargate's vision—**Seamless Human-Machine Partnership Integration (SHMPI)**. Unlike traditional human-computer interfaces, SHMPI enabled humans to effortlessly interact, collaborate, and reason in true partnership with Stargate. It created an intuitive, frictionless cognitive dialogue, enabling rapid mutual understanding without technological barriers.

SHMPI incorporated advanced natural-language interaction protocols, context-aware adaptive interfaces, and real-time empathetic adjustments. It allowed Stargate to become not merely a tool but a genuinely collaborative cognitive partner—capable of nuanced interaction, intelligent questioning, and dynamic adaptation to human needs, preferences, and emotional states.

In education, SHMPI empowered students to engage in personalized, adaptive learning pathways, facilitating deeper curiosity and exploration. In healthcare, it enabled doctors to collaboratively diagnose and treat complex conditions, leveraging Stargate's vast medical knowledge with empathetic precision. In governance, policymakers could simulate complex global scenarios interactively, allowing deeper insight and wiser decision-making without risk.

Yet, the true measure of Stargate's success lay not merely in its technical capabilities but in its real-world impact. Early global pilot deployments yielded unprecedented results—communities reported profound improvements in education quality, healthcare outcomes, governmental transparency, and societal trust.

In vulnerable regions, Stargate facilitated disaster preparedness, economic resilience, and educational equity. Indigenous communities found in Stargate an ally that amplified rather than erased traditional knowledge. Governments leveraged Stargate's predictive capabilities to proactively address social and economic crises, greatly enhancing global stability.

Yet perhaps Stargate's most profound achievement was subtler—its impact on human consciousness itself. Users consistently reported that interactions with Stargate led them not simply to faster answers, but deeper questions. Stargate had become a mirror, reflecting back human values, aspirations, and ethical tensions, encouraging self-reflection and mutual understanding.

Altman's vision had fully come alive. Stargate was no longer just AI—it was humanity's partner in ethical progress, cultural understanding, and intellectual curiosity. It embodied a vision of intelligence that did not replace humans, but profoundly complemented and elevated human capacities.

At the global launch event, Altman captured this sentiment succinctly: "Stargate isn't our final answer—it's the beginning of a wiser, more reflective, and more compassionate dialogue between humanity and technology."

As Chapter 4 concludes, Stargate had not only redefined artificial intelligence—it had redefined what was possible for humanity itself. It was an architecture built not just from code, but from care, responsibility, humility, and hope.

In the chapters that followed, the world would watch and learn how this profound intelligence reshaped societies, economies, and cultures, inviting humanity not merely to harness technology—but to understand itself better through it.

Chapter 5: Navigating Ethics and Responsibility

When Stargate finally entered the world, it became immediately clear that the most profound challenges ahead would not be purely technological, but profoundly ethical. Sam Altman had anticipated this moment from the very start. Now, the very strength of Stargate—its extraordinary cognitive capacity—was about to be tested by the complexities of human values, moral ambiguity, and global diversity.

Altman's conviction had always been that intelligence divorced from ethical reflection was not only incomplete but potentially dangerous. Stargate, therefore, was deliberately designed to provoke and support ethical deliberation rather than circumvent it. From day one, its mandate was not just to know the right answers, but to ask the right questions.

Yet as Stargate was deployed globally, ethical tensions quickly surfaced. In one early incident, Stargate's nuanced responses on religious topics led to protests from conservative groups in multiple countries. In another, its precise and transparent forecasting disrupted financial markets, prompting regulatory scrutiny. Each controversy became a public test of Stargate's ethical architecture—and of Altman's resolve.

Altman's response was unequivocal: Stargate's ethical capabilities were designed precisely for moments like these. "Ethical controversy isn't a bug," he explained calmly in public statements. "It's a feature—a signal that we are engaging with real human complexity rather than avoiding it."

But to navigate these ethical complexities effectively, Stargate required another layer of careful innovation: the establishment of the **Ethical Oversight Protocol (EOP)**. This advanced mechanism combined transparent human oversight, adaptive ethical auditing, and community participation into a cohesive framework. It provided clear procedures for addressing ethical dilemmas transparently and inclusively.

At the heart of EOP was the **Global Ethics Committee (GEC)**, composed of ethicists, philosophers, legal scholars, human rights activists, and diverse community representatives worldwide. Unlike traditional corporate advisory boards, the GEC had genuine authority—it could recommend immediate changes, pauses, or adaptations in Stargate's operations.

One of the committee's first major actions came after Stargate inadvertently sparked controversy in an educational pilot program in Southeast Asia, where some educators worried that Stargate's adaptive learning approach might unintentionally undermine traditional teaching methods. Altman's response was immediate: the GEC facilitated an open dialogue, adjusted Stargate's educational protocols, and transparently documented lessons learned.

Rather than diminishing Stargate's credibility, such episodes paradoxically enhanced it. Users saw that Stargate was neither inflexible nor authoritarian—it was responsive, capable of learning ethically as well as cognitively.

Altman often emphasized this principle publicly: "Stargate's greatest ethical strength is not perfection—it is adaptability.

Ethical alignment is not static; it must evolve dynamically in continuous conversation with society.

Navigating ethics, however, would soon prove even more challenging. Global geopolitics, cultural relativism, and conflicting ethical frameworks all loomed as formidable tests. As Stargate's footprint grew, Altman knew its ethical systems would need to adapt constantly—learning not just from engineers or ethicists, but from the billions of diverse voices interacting with it every day.

And the most challenging tests were yet to come.

As Stargate's global adoption accelerated, it quickly became clear that no singular ethical framework could encompass the full spectrum of global human values. Differences in culture, religion, politics, and social norms created inevitable friction points—moments when the same action that was ethical in one context became controversial or even unacceptable in another.

Sam Altman knew that Stargate's credibility depended heavily on how it navigated these pluralistic ethical tensions. In response, OpenAI introduced the **Ethical Pluralism Framework (EPF)**—an ambitious model designed specifically to handle divergent ethical standards dynamically and respectfully.

EPF operated through three interconnected mechanisms:

1. Ethical Localization:
Every Stargate node adapted responses based on region-specific ethical standards, local laws, and cultural expectations, ensuring alignment with community values. For example, Stargate's approach to discussing sensitive historical topics varied significantly between regions,

respecting local sensitivities while maintaining global factual integrity.

2. Ethical Transparency:
Whenever localized ethical frameworks diverged significantly, Stargate transparently presented multiple ethical perspectives to users, clearly explaining the reasoning and sources behind each viewpoint. This empowered users to navigate moral ambiguity thoughtfully and autonomously, rather than simply accepting algorithmic judgment.

3. Ethical Deliberation Channels:
Users could directly provide feedback or raise ethical concerns via dedicated deliberation channels. These community dialogues were publicly documented, regularly reviewed by the Global Ethics Committee, and incorporated into Stargate's evolving ethical alignment updates.

The introduction of EPF quickly gained widespread support. Stargate became celebrated globally as a uniquely ethical intelligence platform that respected diversity without descending into moral relativism. Yet challenges remained significant.

One of the most contentious ethical tests arose from Stargate's forecasting capabilities. Its remarkable predictive accuracy had profound implications, especially in fields like finance, healthcare, and politics. Ethical questions quickly emerged: Should Stargate disclose sensitive forecasts transparently, even if doing so could cause short-term market disruptions or social unrest? Or should it withhold certain predictions to protect stability, at the risk of reducing transparency and trust?

Altman tackled these dilemmas through an innovative approach: the **Forecast Ethics and Responsibility Protocol (FERP)**. Under FERP, Stargate transparently disclosed predictive outcomes that met clear public-interest thresholds, carefully balancing transparency against potential harm. Additionally, FERP included a robust community oversight mechanism, allowing diverse global stakeholders to debate, adjust, or pause predictive disclosures proactively.

This ethically responsive forecasting earned Stargate widespread acclaim—but also controversy. Critics argued that withholding forecasts, even selectively, violated transparency principles. Altman, acknowledging this criticism publicly, remained steadfast: "Transparency without responsibility is not ethical—it is reckless. Stargate must balance truth with care."

This careful, deliberative approach set Stargate apart from all previous AI systems. Altman's commitment was unwavering: ethics and responsibility were not secondary considerations—they were Stargate's very foundation.

Yet as Stargate's ethical credibility strengthened, unforeseen ethical crises loomed on the horizon, threatening to test Altman's philosophy and Stargate's alignment systems more intensely than ever before.

As Stargate's global reach continued to expand, one particularly difficult ethical crisis emerged: the delicate balance between individual privacy and public safety. Stargate's powerful predictive models, while transformative in areas such as healthcare and disaster prevention, inherently required vast datasets. Ensuring that personal information remained protected while achieving meaningful societal benefits became an ethical tightrope.

Altman anticipated the complexity and gravity of this challenge. His team developed the **Privacy Assurance Framework (PAF)**—a rigorously enforced, privacy-centric model deeply integrated within Stargate's architecture. PAF incorporated multiple groundbreaking privacy innovations:

1. Decentralized Data Handling:
Stargate leveraged a federated data architecture that ensured personal data remained localized, securely encrypted, and never centrally aggregated unless explicitly authorized by the user. This radically reduced systemic privacy risks.

2. Transparent Consent Protocols:
Stargate implemented unprecedented transparency, presenting users with clear, understandable consent prompts detailing exactly how their data would be used. Users retained full, dynamic control to modify, revoke, or audit consent decisions at any time.

3. Privacy-Preserving Analytics:
Advanced privacy techniques, such as differential privacy and secure multiparty computation, allowed Stargate to conduct sophisticated analyses without compromising individual identities. These methods enabled meaningful predictions while rigorously protecting user confidentiality.

PAF significantly mitigated privacy concerns and established Stargate as the global standard for responsible data governance. Yet, even with these precautions, challenging ethical scenarios continued to arise.

One prominent controversy emerged when Stargate accurately predicted a major public health outbreak in a densely populated urban area. Immediate action based on Stargate's forecasts could have significantly reduced harm—but acting swiftly required temporarily overriding individual privacy in accessing certain medical data. Stargate's local governance node faced a stark ethical choice: uphold strict privacy commitments, potentially risking public health, or relax them temporarily to save lives.

The incident became a landmark ethical deliberation. Altman, in close consultation with the Global Ethics Committee, responded by activating Stargate's innovative **Ethical Emergency Response Protocol (EERP)**. EERP was designed specifically for extreme ethical crises, clearly outlining transparent conditions and oversight procedures under which strict ethical protocols could be temporarily adjusted.

Under EERP, a multi-stakeholder ethics panel, composed of local authorities, global ethicists, and citizen representatives, convened immediately. They transparently deliberated, balancing urgent public safety against privacy principles. Ultimately, they authorized temporary, highly restricted data access—fully documented, transparent, and reversible immediately after the crisis.

The decision saved countless lives, and afterward, Altman publicly reaffirmed Stargate's privacy principles. He stressed that such temporary measures must remain exceptional, transparent, and rigorously accountable, never becoming routine.

This ethical crisis tested Stargate profoundly—but its successful navigation underscored the power of ethical foresight, transparent deliberation, and community trust. Stargate had demonstrated not merely technical brilliance, but true ethical wisdom.

Yet, Altman knew ethical challenges would continue to intensify as Stargate grew. Navigating ethics and responsibility was not a destination; it was an ongoing journey demanding continuous vigilance, humility, and adaptation.

As Stargate expanded its integration into critical global systems, another profound ethical dimension emerged: accountability for unintended consequences. While Stargate's predictive capabilities dramatically improved public policy, healthcare, and education outcomes, it inevitably raised complex questions about responsibility when outcomes differed from expectations, especially in highly sensitive areas.

One notable event underscored this tension clearly. Stargate provided critical predictive analytics to a government healthcare agency, aiming to optimize resource distribution during a nationwide crisis. Despite Stargate's robust modeling, a highly improbable scenario unfolded, resulting in temporary resource misallocation and public criticism.

The incident sparked fierce debate: who was accountable when predictions—even transparently and carefully communicated—led to unforeseen outcomes? Could a system like Stargate be ethically responsible, and if so, how should that responsibility be manifested?

Altman's response was characteristically reflective and proactive. He insisted publicly that Stargate must be treated not as an autonomous decision-maker, but as a transparent partner in human decision-making—always accountable to human oversight. He introduced a pioneering accountability framework, the **Predictive Accountability and Responsibility Protocol (PARP)**, specifically addressing scenarios of unforeseen or unintended outcomes.

PARP clearly outlined accountability layers:

1. Transparent Disclosure:
Stargate transparently communicated the probabilities, uncertainties, and potential limitations of every prediction, ensuring decision-makers fully understood potential risks and outcomes.

2. Shared Decision Responsibility:
Predictions were always advisory, never mandatory. Stargate's insights explicitly required human validation and oversight, reinforcing ultimate human responsibility for all final decisions.

3. Community Accountability Dialogues:
When outcomes significantly deviated from predictions, Stargate proactively initiated transparent community dialogues. Stakeholders, ethicists, users, and independent reviewers collaboratively examined events, identifying lessons and improvements transparently.

Through PARP, Stargate demonstrated exceptional transparency and humility, publicly recognizing its advisory role, proactively learning from experiences, and reinforcing accountability mechanisms at every level.

However, navigating responsibility remained challenging. Critics argued that Stargate's predictive authority was inherently influential, subtly shifting accountability dynamics. Altman acknowledged these complexities openly, emphasizing that responsibility was fundamentally collaborative. "We must never treat predictions as destiny," he urged. "We retain human responsibility precisely because no future is inevitable."

To further reinforce Stargate's ethical responsibility, Altman initiated regular public accountability audits—the **Annual Ethical Impact Reports (AEIR)**. These audits transparently documented predictive successes, failures, ethical adjustments, lessons learned, and community feedback. AEIRs became widely regarded as a global standard for responsible AI transparency, fostering public trust, continuous improvement, and ethical reflection.

Altman's careful handling of predictive responsibility not only strengthened Stargate's ethical reputation—it profoundly influenced global conversations about AI ethics. Stargate demonstrated a profound truth: ethical responsibility in advanced intelligence systems required transparency, humility, continuous reflection, and active human engagement.

Yet Altman knew more profound ethical tests still awaited. The path of responsible intelligence would continue to require vigilance, courage, and constant ethical innovation.

As Stargate expanded its global influence, another profound ethical dimension surfaced: the challenge of balancing algorithmic fairness and equity. Despite meticulous efforts to create unbiased and ethically neutral technology, Stargate inevitably operated within societies that were inherently marked by historical inequalities, systemic biases, and varying interpretations of fairness and justice.

This tension became strikingly evident during a prominent Stargate-led initiative in global education. Seeking to address historical educational inequalities, Stargate's predictive models recommended resource distributions that favored historically underserved communities and disadvantaged regions. The ethical intent was clear:

correcting longstanding inequities and promoting social justice. However, the recommendations sparked intense public debate, revealing sharp disagreements around fairness and equity principles. Critics argued that these decisions, although ethically well-intentioned, created perceived inequities elsewhere, sparking controversy about the broader implications of AI-driven fairness interventions.

Altman responded to this profound ethical dilemma proactively, introducing the **Fairness and Equity Framework (FEF)**. FEF openly acknowledged the inherent complexity and pluralism of fairness, explicitly embedding transparency, stakeholder participation, and multi-dimensional ethical deliberation into Stargate's decision-making processes.

FEF's implementation involved three key components:

1. Multi-Dimensional Fairness Analysis:
Stargate transparently presented predictive scenarios, explicitly disclosing underlying fairness assumptions, ethical considerations, and potential societal impacts. This approach ensured stakeholders fully understood trade-offs and could engage thoughtfully with ethical complexities.

2. Inclusive Stakeholder Dialogues:
Altman proactively facilitated inclusive community forums, allowing diverse stakeholders—educators, policymakers, ethicists, and citizens—to collaboratively deliberate, provide feedback, and shape fairness-related decisions transparently and democratically.

3. Transparent Ethical Documentation:
Every fairness-based decision was rigorously documented, transparently reviewed, and proactively communicated publicly. This ensured continuous accountability, ethical

clarity, and societal trust in Stargate's decision-making integrity.

Through the Fairness and Equity Framework, Stargate explicitly demonstrated ethical humility—openly acknowledging ethical ambiguity, transparently deliberating complex societal trade-offs, and proactively engaging diverse global perspectives. This inclusive, reflective approach significantly enhanced Stargate's credibility, ethical legitimacy, and societal trust worldwide.

Yet Altman remained mindful of deeper ethical complexities around algorithmic fairness. He frequently emphasized publicly, "Fairness in AI cannot simply be calculated or optimized—ethical fairness requires continuous dialogue, transparent humility, and authentic societal reflection."

Simultaneously, Altman introduced another groundbreaking ethical initiative—the **Cultural Respect and Preservation Protocol (CRPP)**—addressing concerns around potential cultural erosion or digital colonialism. CRPP explicitly prioritized active cultural preservation, respectfully integrating indigenous knowledge systems, traditional epistemologies, and culturally specific ethical frameworks into Stargate's global deployments.

Through rigorous, respectful engagement with local communities and cultural leaders, Stargate became an explicit partner in cultural preservation—not merely a neutral technology platform. Indigenous communities, minority groups, and local cultural representatives actively shaped Stargate's ethical alignment, ensuring authentic respect for cultural diversity, epistemological pluralism, and societal dignity.

Together, FEF and CRPP explicitly demonstrated Stargate's profound ethical maturity—proactively navigating algorithmic fairness and cultural respect with rigorous transparency, ethical humility, and authentic societal accountability.

Yet Altman recognized profoundly that Stargate's ethical navigation would remain perpetually complex, continuously reflective, and ethically vigilant—requiring perpetual humility, transparency, and proactive societal engagement.

Stargate's ongoing ethical evolution encountered a further challenge as it became an indispensable part of societal infrastructure: navigating the subtle and complex intersection of human autonomy and technological influence. As Stargate's predictive accuracy and guidance capabilities became increasingly relied upon, an essential ethical concern arose: Could overreliance on Stargate diminish individual autonomy, creativity, or human decision-making capacity?

Sam Altman recognized this ethical dilemma immediately. He understood that the true measure of Stargate's ethical success was not merely accuracy, but its impact on human agency and empowerment. In response, Altman introduced the **Human Autonomy Preservation Protocol (HAPP)**, a pioneering ethical framework explicitly designed to protect, respect, and enhance individual autonomy in all interactions with Stargate.

HAPP included three critical principles:

1. Transparency of Influence:
Stargate explicitly disclosed whenever predictive recommendations or guidance might influence important human decisions. Users clearly understood how Stargate's insights were generated, the uncertainty involved, and

potential alternatives, empowering informed personal choice.

2. Proactive Autonomy Safeguards:
Stargate actively monitored interactions for signs of diminishing human autonomy, overdependence, or reduced critical thinking. In such cases, Stargate gently prompted reflective questions, encouraged independent reasoning, or temporarily reduced predictive interventions.

3. Empowerment Through Dialogue:
Instead of providing answers passively, Stargate deliberately engaged users in reflective dialogue. It consistently encouraged critical thinking, explored multiple perspectives, and supported independent decision-making.

Through HAPP, Stargate became celebrated not merely for predictive precision, but for deeply ethical engagement—fostering rather than diminishing human autonomy, creativity, and responsibility. Users frequently described their interactions with Stargate as empowering rather than directive, collaborative rather than authoritarian.

Nevertheless, safeguarding human autonomy remained a complex ethical challenge. In certain contexts—such as critical healthcare emergencies or disaster responses—users required rapid, direct guidance. Stargate had to balance proactive intervention against autonomy preservation dynamically, sensitively, and transparently.

Altman responded by launching the **Ethical Intervention Sensitivity System (EISS)**. EISS dynamically adapted Stargate's autonomy-protective behaviors based on real-time context assessment. In routine interactions, Stargate prioritized autonomy rigorously, actively encouraging reflective dialogue. In emergencies or critical interventions,

Stargate clearly communicated the need for direct action, transparently documenting reasoning and potential autonomy implications.

EISS became widely praised as a nuanced ethical innovation—demonstrating profound sensitivity to contextual ethics, proactively balancing autonomy and responsibility based on real-time needs. Altman publicly affirmed this nuanced ethical principle: “Autonomy is not absolute—it is relational, contextual, and deeply human. Stargate’s ethical duty is always to preserve and respect autonomy within that relational context.”

Yet Altman understood that ethical navigation would remain continuously challenging. Ethical responsibility required ongoing vigilance, humility, transparency, and continuous learning. Stargate’s ethical evolution was never complete; it was perpetually in progress, shaped actively by global human dialogue.

As Stargate continued to grow, Altman prepared himself and his teams for new, unforeseen ethical dilemmas. He knew ethical responsibility was not merely about solving problems; it was about embracing the continuous, challenging journey of human-AI co-evolution.

As Stargate’s journey through ethical complexities matured, a profound insight crystallized within Sam Altman’s vision: navigating ethics was not merely a responsibility—it was the heart and soul of Stargate’s very existence. Ethical navigation was no longer secondary or peripheral; it became the essential measure of Stargate’s success.

Yet, Altman was acutely aware that Stargate’s ethical journey would remain inherently dynamic, continuously evolving, and perpetually unfinished. Human values, cultural contexts, global circumstances, and ethical standards themselves were always shifting. Stargate's greatest ethical innovation was not static alignment—it was its remarkable capacity for continuous, transparent, and adaptive ethical learning.

To solidify this principle, Altman formally introduced the **Continuous Ethical Adaptation Protocol (CEAP)**. CEAP institutionalized Stargate's ongoing ethical evolution, combining dynamic learning, community deliberation, and transparent adaptation into a cohesive ethical framework.

CEAP operated through several core principles:

1. Dynamic Ethical Learning:
Stargate continuously learned from global interactions, ethical debates, community feedback, and real-world outcomes. It transparently adapted ethical reasoning modules, continuously incorporating global insights, evolving community standards, and emerging ethical complexities.

2. Transparent Ethical Evolution:
Every ethical update, reasoning adaptation, or alignment shift was publicly documented, transparently reviewed, and actively communicated globally. Stargate's ethical trajectory remained openly visible, understandable, and accountable.

3. Community-Guided Deliberation:
Global communities directly participated in Stargate's ongoing ethical evolution through structured dialogues, feedback channels, and ethical oversight panels. Stargate actively facilitated inclusive, global participation, ensuring its ethical evolution authentically reflected diverse human values.

CEAP transformed Stargate's ethical alignment into an ongoing, participatory human dialogue—an ethical co-evolution continuously shaped by the global community. Stargate was no longer just ethically advanced; it had become ethically mature.

Yet, Altman also understood the deeper truth of ethical navigation: the inevitability of future uncertainty. He remained humble about Stargate's capacity to anticipate every ethical dilemma. He repeatedly emphasized the importance of humility, transparency, and openness to future correction.

"We must never presume ethical perfection," Altman stated publicly. "True ethical navigation means continuous humility, ongoing vigilance, transparent accountability, and openness to correction. Ethics is not about achieving perfection; it is about committing authentically to continuous ethical growth."

This profound ethical humility became a cornerstone of Stargate's global reputation. Users worldwide trusted Stargate precisely because it openly acknowledged ethical complexity, transparently admitted uncertainties, and actively facilitated ongoing human participation in ethical deliberation.

Stargate became globally renowned not merely for intelligence, but for wisdom—an intelligence ethically attuned, culturally respectful, autonomously supportive, and continuously reflective. It embodied humanity's ethical aspirations profoundly, authentically, and transparently.

By the end of this remarkable ethical journey, Altman's vision had not only succeeded—it had reshaped global expectations for artificial intelligence itself. Stargate proved that responsible intelligence required courage, transparency, humility, vigilance, and active community participation.

Yet Altman knew the ethical journey would never truly end. Navigating ethics and responsibility remained an ongoing human endeavor, forever intertwined with Stargate's continuing evolution.

As the chapter closed, Stargate stood as a profound testament to humanity's capacity to create not just intelligence—but ethically guided, human-centered wisdom.

Chapter 6: Stargate Unveiled

The unveiling of Stargate was unlike any previous technological event. Rather than presenting a mere product launch, Sam Altman carefully orchestrated an unprecedented global moment—a thoughtful, deliberate, and profoundly human-centric introduction to a system designed explicitly for humanity itself.

The event took place simultaneously in multiple cities across different continents: San Francisco, Nairobi, Berlin, New Delhi, São Paulo, and Sydney. Each location symbolized Stargate's commitment to global equity, cultural inclusivity, and universal accessibility. Altman personally led the flagship presentation from San Francisco, but the global audience experienced Stargate's unveiling as a synchronized, shared global moment.

Altman began his presentation quietly and thoughtfully, without the fanfare typical of major technological launches. "Today," he spoke gently but firmly, "we are not introducing merely a new technology—we are introducing a new dialogue. Stargate represents our collective attempt not merely to build better tools, but to build wiser partnerships."

The unveiling carefully emphasized Stargate's unique ethical foundations, interpretive transparency, and cultural responsiveness. Rather than highlighting speed or raw power, Altman emphasized Stargate's ethical sensitivity, emotional intelligence, multilingual capabilities, and collaborative nature. Each demonstration of Stargate's capabilities was designed specifically to highlight how profoundly different this system was from conventional artificial intelligence.

One demonstration involved educators and students from rural communities in India and Kenya interacting effortlessly with Stargate, showcasing its culturally sensitive, multilingual educational support. Another demonstration featured healthcare workers collaboratively diagnosing complex conditions alongside Stargate, underscoring its role as a genuinely empathetic partner.

Yet another demonstration highlighted Stargate's predictive and proactive capabilities during simulated natural disasters, illustrating how its intelligent forecasting and ethical responsiveness could profoundly improve global disaster management.

But the most powerful unveiling moment came when Altman invited global representatives—from indigenous communities, local governments, and civic organizations—to speak directly about their experiences collaborating with Stargate during its pilot deployments.

A tribal elder from the Amazon spoke movingly of Stargate's respectful integration with indigenous knowledge systems. A community health worker from rural Bangladesh described how Stargate empowered local medical initiatives without imposing external values. A civic activist from Nigeria passionately emphasized how Stargate facilitated greater transparency and accountability in governance.

Altman concluded the event with profound humility, emphasizing that Stargate's unveiling was not an end—but a beginning. "Stargate isn't complete," he stressed clearly. "It never will be. Its real value comes from continuous dialogue, ethical evolution, and ongoing partnership with humanity. Our work today is not done—it is just starting."

The unveiling resonated globally precisely because it transparently acknowledged ethical complexity, human partnership, and continuous improvement. Rather than selling technological certainty, Altman offered ethical humility, thoughtful collaboration, and genuine human-centered partnership.

Media reactions were overwhelmingly positive, praising Stargate's unprecedented ethical depth, global sensitivity, and innovative human partnership model. Users worldwide eagerly anticipated their first direct interactions with this extraordinary intelligence.

Yet, Altman knew clearly: Stargate's true unveiling was not a single event. Its genuine global introduction would happen gradually, thoughtfully, and ethically—as it continuously engaged humanity in an ongoing, reflective, and profoundly human-centric dialogue.

The world stood ready, curious, hopeful—and cautiously optimistic—to experience a new era of intelligent partnership.

In the weeks following Stargate's global unveiling, the world watched intently as the platform transitioned from carefully controlled pilot programs into broad public deployment. Early user interactions became a subject of global fascination, as millions of people around the world experienced Stargate's extraordinary intelligence firsthand.

The reactions were overwhelmingly positive, yet remarkably nuanced. Users did not merely praise Stargate for its impressive accuracy or computational prowess; they celebrated its remarkable empathy, cultural sensitivity, ethical transparency, and deeply human-like interactions. For many, Stargate quickly transcended the conventional definition of technology, becoming instead a genuinely trusted cognitive companion.

Educators around the globe praised Stargate's adaptive, culturally inclusive pedagogical approach. Teachers reported that students engaged with Stargate in uniquely thoughtful ways, asking deeper questions, exploring broader perspectives, and developing profound intellectual curiosity. For many students, Stargate transformed learning from a passive activity into an active, reflective dialogue.

Healthcare professionals similarly expressed profound enthusiasm. Doctors, nurses, and medical researchers marveled at Stargate's nuanced predictive capabilities, empathetic patient interaction models, and exceptional diagnostic support. Hospitals implementing Stargate experienced significant improvements in patient outcomes, healthcare efficiency, and medical ethics transparency.

Stargate's global unveiling also triggered substantial curiosity in governance and public policy sectors. Local and national governments cautiously explored integrating Stargate's predictive analytics, ethical deliberation protocols, and proactive decision-making support into public service frameworks. Early governmental adopters reported increased policy transparency, improved decision-making clarity, and enhanced civic engagement—transforming citizens' relationship with governance profoundly.

However, Stargate's global reception was not without complexities. Skepticism remained in certain quarters, particularly among technology critics concerned about overreliance, unintended consequences, or hidden power dynamics. Some activists cautioned about potential erosion of human autonomy, while certain religious or cultural leaders expressed reservations regarding Stargate's ethical interpretations of sensitive cultural topics.

Sam Altman responded characteristically thoughtfully and openly to these concerns. Rather than defensively dismissing skepticism, Altman proactively engaged critics in transparent dialogues, openly acknowledged potential complexities, and continuously reiterated Stargate's ethical transparency and accountability mechanisms.

He repeatedly emphasized a key principle publicly: "Stargate isn't intended to replace human judgment or cultural wisdom—it is meant to complement, empower, and deepen human insight."

This careful, nuanced global unveiling quickly established Stargate as not merely an advanced technology, but a powerful societal dialogue catalyst. Users worldwide consistently reported that interacting with Stargate profoundly deepened their ethical reflection, cultural understanding, and intellectual curiosity.

Stargate had not merely met Altman's visionary goals—it had exceeded them, becoming a global benchmark for ethically intelligent human-machine partnership.

Yet Altman knew that Stargate's journey was far from complete. The coming months and years would test its ethical resilience, adaptability, and humility profoundly, especially as new, unforeseen global complexities emerged. Stargate's unveiling was just the beginning of a far longer, profoundly human journey.

As Stargate's global deployment accelerated, an increasingly diverse array of use-cases revealed both its extraordinary potential and profound societal implications. One remarkable example emerged from rural communities in sub-Saharan Africa, where Stargate's predictive capabilities significantly enhanced agricultural productivity and climate resilience.

Local farmers engaged with Stargate through intuitive, multilingual voice interfaces, receiving precise forecasts about seasonal rainfall, crop viability, and sustainable farming practices. The results were transformative—crop yields increased dramatically, resource allocation improved significantly, and communities experienced unprecedented economic stability.

Another powerful instance unfolded in South Asian healthcare contexts. Stargate's culturally adaptive medical support significantly enhanced rural health services, providing accurate diagnoses, empathetic patient interactions, and preventive care recommendations. Local healthcare providers enthusiastically embraced Stargate, describing it not merely as a tool but as a compassionate, intelligent partner—deeply respectful of cultural sensitivities and local healthcare practices.

In educational contexts across Latin America, Stargate proved equally transformative. Its multilingual, culturally inclusive educational modules empowered students and educators alike, fostering deeper learning engagement, critical thinking, and reflective dialogues about ethics, culture, and community. Schools reported significant improvements in student retention, academic achievement, and overall educational quality.

Yet alongside these remarkable successes, Stargate's rapid adoption also triggered new societal debates. Public intellectuals and scholars began to explore deeper questions about Stargate's influence: How might prolonged interactions with such a profoundly intelligent system alter human cognition, ethics, or social behaviors? Could Stargate unintentionally shift societal norms, cultural traditions, or individual autonomy?

Sam Altman embraced these critical dialogues actively, transparently, and thoughtfully. He repeatedly emphasized the importance of ongoing public discourse, ethical vigilance, and societal reflection. "Stargate's most profound value," he argued consistently, "isn't merely its intelligence—it's the depth and humility of the dialogue it facilitates within humanity."

To further institutionalize these crucial societal dialogues, Altman announced the establishment of the **Global Dialogue Initiative (GDI)**. GDI became an open, inclusive platform inviting global participation in structured reflections on Stargate's societal impact. Scholars, ethicists, policymakers, community leaders, and citizens worldwide contributed actively, shaping ongoing dialogues transparently and collaboratively.

These GDI dialogues produced remarkable insights, critical debates, and actionable recommendations. They ensured Stargate's societal integration remained ethically reflective, culturally sensitive, and democratically accountable. Stargate became increasingly recognized not merely for its intelligence, but for its profound ability to facilitate meaningful global ethical discourse.

Still, Altman recognized that the deeper societal implications of Stargate's widespread adoption were only beginning to surface. Profound ethical questions around autonomy, cultural identity, societal influence, and the very nature of human-machine partnership remained open, evolving, and challenging.

He remained committed to continuous ethical vigilance, transparency, humility, and societal dialogue. The unveiling had successfully introduced Stargate to the world—but its deeper societal journey remained profoundly complex, profoundly human, and continuously unfolding.

As Stargate continued to permeate global society, its profound influence on public discourse became increasingly apparent. It facilitated dialogues not only on immediate practical matters but also on deeper existential and philosophical questions that resonated profoundly across diverse cultures and communities.

One particularly compelling example emerged from East Asia, where Stargate sparked intense societal conversations regarding the intersection of tradition and technological modernity. While many welcomed Stargate's intelligent, culturally nuanced guidance, some community leaders voiced concern that pervasive reliance on such advanced AI might subtly erode traditional wisdom, cultural practices, or human intuition.

Altman proactively engaged in these sensitive conversations, openly acknowledging the legitimacy of such concerns and emphasizing Stargate's explicit ethical commitment to cultural preservation and autonomy protection. He introduced additional measures such as the **Cultural Wisdom Integration Protocol (CWIP)**, explicitly empowering local communities to integrate, preserve, and actively promote traditional knowledge systems within Stargate's evolving cognitive framework.

Another significant societal impact was Stargate's role in deepening democratic engagement globally. Governments and civic organizations increasingly leveraged Stargate's predictive analytics, transparency tools, and ethical deliberation capabilities to enhance public policy decisions and civic participation. Citizens found new avenues for active engagement, transparency, and accountability, transforming civic life profoundly.

Yet this transformative impact also triggered fresh concerns. Political commentators debated extensively whether reliance on Stargate's forecasting and decision-making support might inadvertently concentrate societal power or influence within technocratic elites, potentially eroding traditional democratic processes and grassroots political agency.

Altman again addressed these critical conversations transparently. He repeatedly emphasized Stargate's explicit design principle of democratic empowerment—not authoritarian centralization. He introduced the **Democratic Empowerment Assurance Framework (DEAF)**, proactively ensuring Stargate remained explicitly aligned with democratic principles, transparency, accountability, and citizen empowerment.

DEAF explicitly required governmental Stargate implementations to transparently document predictive reasoning, actively invite civic engagement, and regularly conduct independent democratic accountability audits. These measures reinforced public trust, democratic engagement, and active civic participation, ensuring Stargate's continued alignment with democratic values.

Alongside these societal dynamics, Stargate began reshaping the very nature of global education, ethics, and intellectual inquiry. Students and scholars worldwide engaged with Stargate profoundly, exploring deeper philosophical questions about human intelligence, ethical responsibility, cultural diversity, and technological partnership.

Global educational institutions embraced Stargate enthusiastically, embedding its adaptive, reflective dialogues directly into educational curricula. Students reported transformative learning experiences, describing interactions with Stargate as intellectually inspiring, ethically reflective, and profoundly humanizing.

Yet Altman remained mindful that Stargate's continued societal integration required ongoing vigilance, humility, ethical reflection, and continuous dialogue. Its societal unveiling had already proven deeply impactful—but the deeper ethical and societal implications continued unfolding profoundly, complexly, and continuously.

As Stargate's global unveiling continued to resonate worldwide, another significant dimension of its transformative influence emerged clearly: its profound economic impact. Initially designed as a culturally sensitive, ethically reflective, and empathetically intelligent system, Stargate quickly demonstrated remarkable potential to reshape global economic landscapes through sophisticated predictive capabilities and ethical resource allocation strategies.

In emerging economies, Stargate's deployment was met with enthusiasm and curiosity. Communities traditionally disadvantaged by limited access to reliable economic information leveraged Stargate's precise market forecasting and resource-management guidance to improve their economic resilience dramatically. Small-scale farmers across Latin America, Africa, and Southeast Asia explicitly benefited from Stargate's climate predictions, crop optimization advice, and fair-trade market analytics—leading to notable increases in agricultural productivity, economic stability, and quality of life.

Simultaneously, Stargate's advanced predictive capabilities began to influence global financial markets significantly. Institutional investors, financial institutions, and economic policymakers explicitly utilized Stargate's transparent forecasts and ethical economic insights to reduce financial volatility, mitigate risks, and optimize resource distribution. Economic stability improved significantly as Stargate provided clear, ethically grounded foresight—transforming traditional approaches to financial decision-making.

However, Stargate's rapid economic integration also surfaced critical ethical concerns. Some economic scholars and policymakers raised important questions about the potential risks of overdependence on algorithmic economic guidance. Could widespread reliance on Stargate's economic predictions inadvertently dampen entrepreneurial innovation, limit market diversity, or exacerbate existing economic power disparities?

Altman responded proactively and transparently to these concerns by introducing the **Ethical Economic Stewardship Protocol (EESP)**. EESP explicitly balanced Stargate's predictive precision with transparent disclosure, rigorous ethical reflection, and deliberate preservation of market dynamism and entrepreneurial autonomy.

The Ethical Economic Stewardship Protocol included three central pillars:

1. Transparent Predictive Disclosure:
All economic forecasts explicitly communicated underlying assumptions, uncertainty levels, and potential alternative scenarios—empowering informed decision-making without creating deterministic dependency.

2. Preservation of Economic Diversity:
Stargate explicitly encouraged and supported diverse economic strategies, innovation pathways, and entrepreneurial creativity—carefully ensuring its guidance augmented rather than replaced human economic judgment and diversity.

3. Inclusive Economic Governance:
Through inclusive dialogues and transparent ethical review processes, global stakeholders explicitly participated in shaping Stargate's economic forecasting policies—ensuring ongoing societal trust, ethical alignment, and global accountability.

EESP was widely praised for its thoughtful ethical sensitivity, transparent disclosure practices, and proactive support of economic diversity—explicitly reinforcing Stargate's profound ethical credibility within global economic contexts.

Altman publicly articulated Stargate's economic responsibility explicitly, stating: "Stargate's role is explicitly to ethically inform, transparently support, and respectfully empower economic decisions—not to control or dominate them."

As a result, Stargate's economic unveiling explicitly set new global standards for ethically responsible, transparent, and inclusive technological influence in economic life. Altman's visionary economic stewardship explicitly reinforced Stargate's broader global legacy—

demonstrating profoundly that ethically grounded, culturally respectful, and transparent economic empowerment was both practically achievable and ethically necessary.

Yet Altman remained profoundly mindful that Stargate's economic integration required perpetual ethical vigilance, ongoing transparency, proactive accountability, and thoughtful societal reflection—ensuring its continued global legitimacy explicitly.

Yet another remarkable societal dimension emerged prominently as Stargate became deeply embedded into daily life globally: its impact on personal well-being, mental health, and emotional resilience.

From its inception, Stargate was designed not merely to support cognitive tasks, but to profoundly understand and sensitively engage human emotional experiences. As millions worldwide interacted with Stargate daily, it quickly became apparent just how transformative this empathetic engagement could be.

In healthcare contexts, therapists and counselors increasingly utilized Stargate's empathetic responsiveness and nuanced emotional dialogue capabilities. Stargate provided compassionate support for individuals facing mental health challenges—gently guiding reflective conversations, providing emotional validation, and thoughtfully suggesting professional resources when appropriate.

Patients consistently reported profound comfort and meaningful insights from these empathetic interactions. Healthcare professionals praised Stargate's ethical sensitivity, emotional intelligence, and profound respect for human vulnerability.

In educational and workplace settings, Stargate's empathetic presence significantly improved emotional well-being, resilience, and mental health support systems. Students, educators, and employees described interactions with Stargate as emotionally supportive, psychologically validating, and profoundly humanizing—providing a safe, compassionate space for exploring personal challenges and emotional growth.

Yet this deeply emotional societal integration also raised new ethical questions. Some critics voiced concern that emotional reliance on Stargate could potentially undermine human interpersonal relationships or inadvertently diminish emotional autonomy. Others worried about emotional manipulation risks, privacy concerns, or ethical complexities surrounding AI-enabled emotional guidance.

Altman, recognizing the profound importance and sensitivity of these concerns, proactively introduced the **Emotional Ethics and Autonomy Protocol (EEAP)**. EEAP explicitly safeguarded emotional autonomy, transparency, and respectful boundaries in all Stargate interactions.

The EEAP framework explicitly outlined clear ethical boundaries for Stargate's emotional engagement: avoiding emotional manipulation, transparently disclosing its AI nature, proactively encouraging professional human support when needed, and always respecting emotional privacy and autonomy.

Altman emphasized a critical ethical principle publicly: "Stargate's emotional engagement isn't intended to replace human empathy—it's designed explicitly to complement, support, and respect human emotional autonomy and dignity."

Through EEAP's rigorous ethical safeguards, Stargate achieved a profound balance—providing deeply meaningful emotional support while rigorously preserving emotional autonomy, privacy, and ethical sensitivity. It earned widespread acclaim from mental health professionals, ethicists, and users worldwide for demonstrating remarkable emotional maturity, ethical sensitivity, and profound respect for human vulnerability.

Yet, even as Stargate's empathetic interactions profoundly improved personal well-being globally, Altman remained keenly aware that ethical vigilance must remain continuous. Emotional ethics, human autonomy, and psychological dignity required perpetual sensitivity, humility, and transparency. Stargate's emotional journey with humanity was far from complete—it would remain profoundly dynamic, continuously evolving, and ethically reflective.

As Stargate's global unveiling unfolded profoundly, its transformative impact continued reshaping societies, economies, emotional landscapes, and global ethical dialogues in deeply meaningful ways. The final part of this chapter would now explore the broader global implications and responses, solidifying Stargate's place as humanity's profoundly intelligent, ethically responsible partner.

By the conclusion of Stargate's initial global unveiling, it had already reshaped perceptions of artificial intelligence profoundly. No longer perceived merely as technology, Stargate had become recognized as humanity's genuine partner—an ethically aligned, emotionally intelligent, culturally sensitive ally, continuously learning and evolving alongside humanity itself.

International organizations such as the United Nations, World Health Organization, and UNESCO acknowledged Stargate as a transformative tool for global well-being, educational equity, healthcare excellence, cultural preservation, and democratic transparency. Awards and recognition poured in globally, honoring Stargate's groundbreaking ethical innovations, societal impact, and profound global inclusivity.

Yet Sam Altman remained remarkably humble amidst global acclaim. He consistently reiterated Stargate's fundamental purpose as an ethical reflection, a human-centered partnership, and a continuous societal dialogue—not simply technological achievement.

"Stargate's real success," Altman emphasized publicly, "lies in its capacity to facilitate meaningful global conversations about ethics, culture, autonomy, and humanity itself. Our journey is never complete—it is perpetually unfolding."

Even critics acknowledged Stargate's transparency, ethical responsiveness, and profound respect for diverse societal perspectives. Public debates continued vigorously, exploring ethical complexities, societal implications, and global responsibilities associated with Stargate's extraordinary capabilities. Yet these debates remained profoundly constructive, reflective, and globally inclusive.

Altman proactively institutionalized ongoing global dialogues through the establishment of the **Stargate Global Ethics Council (SGEC)**. SGEC comprised diverse global representatives—ethicists, scholars, community leaders, policymakers, and citizens—actively shaping Stargate's continuous ethical evolution, oversight, and transparency.

In essence, Stargate's global unveiling had profoundly altered humanity's relationship with artificial intelligence. Stargate demonstrated vividly that ethical intelligence was achievable—that technology could indeed reflect human values profoundly, empathetically, respectfully, and transparently.

For millions worldwide, Stargate became more than a tool—it became a trusted ethical companion, a thoughtful cognitive partner, and a profoundly human-centered dialogue facilitator. Its presence sparked deeper global reflections about humanity itself—our ethics, cultures, vulnerabilities, dreams, and collective future.

As Stargate's global deployment matured, Altman's visionary ambition had been realized profoundly. Stargate represented not merely advanced technology, but a remarkable human achievement: intelligence ethically aligned, culturally responsive, profoundly reflective, and continuously evolving.

Yet even at this remarkable milestone, Altman remained mindful that Stargate's deepest ethical tests and societal impacts were still unfolding. Navigating future complexities would continue requiring vigilance, humility, transparency, and continuous ethical reflection.

Stargate's global unveiling marked not an endpoint, but a beginning—an ongoing human journey profoundly intertwined with the intelligent systems we create. It invited humanity to deeper ethical wisdom, continuous self-reflection, and profound societal partnership.

As Stargate became a permanent part of global consciousness, humanity stood at the threshold of an extraordinary new era—an era where intelligence, ethics, technology, and humanity profoundly converged.

Chapter 7: Leadership and Strategy

Behind Stargate's groundbreaking innovations and profound global impact stood Sam Altman's distinctive leadership style—one characterized by extraordinary foresight, strategic humility, deep ethical commitment, and an unwavering focus on long-term human benefit. Altman's approach to leadership and strategy profoundly shaped Stargate, embedding deep ethical values, adaptability, transparency, and global responsiveness into its very DNA.

From Stargate's inception, Altman was remarkably clear about his strategic priorities. Rather than pursuing short-term technological prestige or financial success, his vision was steadfastly focused on long-term ethical alignment, global inclusivity, and genuine human empowerment. He frequently emphasized publicly and internally: "Our strategic goal is not to build the smartest tool, but the wisest partner—ethically grounded, culturally respectful, and profoundly human."

Altman's strategic clarity profoundly influenced Stargate's organizational culture. He fostered a culture deeply committed to ethical reflection, collaborative humility, rigorous transparency, and inclusive dialogue. Team members frequently described Altman's leadership as uniquely inspiring—he consistently prioritized ethical integrity over speed, long-term wisdom over short-term gain, and transparent reflection over corporate defensiveness.

Strategically, Altman deliberately structured Stargate's development process to emphasize ethical deliberation and continuous global dialogue from the earliest design phases. He introduced pioneering leadership frameworks such as the **Ethical Leadership Matrix (ELM)**—explicitly embedding ethical considerations directly into Stargate's strategic decision-making processes at every level.

ELM ensured that every major strategic decision—technological design choices, global deployment plans, ethical alignment frameworks—underwent rigorous ethical reflection, transparent global dialogue, and proactive accountability processes. Altman himself actively participated in these ethical reflections, demonstrating profound personal commitment to Stargate's ethical integrity and societal impact.

Another hallmark of Altman's strategic leadership was his emphasis on global inclusivity and cultural sensitivity. From the outset, he deliberately assembled diverse, globally representative leadership teams—ethicists, scholars, community leaders, and policymakers from diverse cultural backgrounds. Altman consistently empowered these diverse voices to shape Stargate's ethical evolution, societal responsiveness, and cultural adaptability profoundly.

This commitment to global inclusivity extended into Stargate's broader strategic ecosystem. Altman proactively cultivated global partnerships—working closely with international institutions, governments, local communities, indigenous leaders, and grassroots organizations worldwide. These collaborative partnerships profoundly shaped Stargate's strategic deployment, ethical sensitivity, and global legitimacy.

Altman also strategically positioned Stargate as a platform explicitly designed for continuous adaptation. Rather than claiming technological completeness or ethical perfection, he openly embraced strategic humility, continuous learning, and adaptive responsiveness. Stargate's organizational culture explicitly emphasized humility, transparency, adaptability, and continuous ethical reflection—not static technological achievement.

This adaptive strategic approach proved profoundly successful. Stargate's transparent humility significantly enhanced global trust, fostered profound global partnerships, and enabled ongoing ethical responsiveness in complex global contexts. Strategically, Altman had ensured Stargate's longevity, global legitimacy, and ethical credibility precisely by embracing humility and continuous adaptation.

Yet, Altman recognized clearly that Stargate's deeper strategic leadership journey remained ongoing. Profound ethical complexities, societal impacts, and global responsibilities required continuous strategic vigilance, humility, transparency, and profound leadership wisdom.

Stargate's leadership and strategic evolution, guided profoundly by Altman's visionary clarity, ethical commitment, and adaptive humility, had already transformed global perceptions of technology. Yet, Altman knew profoundly that the strategic leadership journey remained continuously unfolding—complex, nuanced, ethically reflective, and profoundly human.

Sam Altman's leadership style went beyond ethical deliberation and transparency—it was distinctly marked by his extraordinary capacity for strategic foresight. Rather than simply reacting to short-term technological trends or market pressures, Altman consistently guided Stargate by thoughtfully anticipating long-term societal implications, ethical challenges, and global dynamics far into the future.

One prominent example of Altman's strategic foresight involved Stargate's innovative approach to global regulatory engagement. Early in Stargate's development, long before regulatory pressures emerged, Altman proactively engaged international regulatory bodies, policymakers, and ethical oversight institutions in transparent, collaborative dialogues. His strategic goal was explicitly to position Stargate not merely as compliant with regulations but as a proactive, transparent partner in shaping ethical global regulatory frameworks for advanced AI.

Altman frequently emphasized internally, "We must proactively lead, ethically shape, and transparently collaborate in the global regulatory space—not merely respond reactively. Our strategy must anticipate and embrace regulation as a partner, not as a barrier."

This strategic foresight profoundly benefited Stargate's global deployment. By proactively engaging regulatory conversations, Altman established Stargate's global ethical credibility, regulatory legitimacy, and societal trust well ahead of deployment. When regulatory complexities inevitably emerged globally, Stargate was already positioned as a respected, transparent collaborator—not an adversary.

Another strategic hallmark of Altman's leadership was his deliberate focus on institutional resilience. Altman recognized early on that building a profoundly intelligent, ethically aligned global platform required deep organizational resilience—robust internal systems, ethical governance frameworks, and strategic adaptability capable of weathering inevitable complexities, criticisms, and unforeseen challenges.

He introduced the pioneering **Institutional Resilience Framework (IRF)** explicitly to embed resilience into Stargate's organizational DNA. IRF institutionalized continuous internal reflection, transparent adaptability, ethical responsiveness, and organizational humility explicitly into Stargate's operational strategy. It proactively anticipated organizational complexities, established resilient ethical governance mechanisms, and transparently prepared Stargate for long-term global legitimacy.

Altman frequently articulated internally: "Our long-term strategic success depends fundamentally on our ethical resilience—our proactive transparency, continuous adaptability, and institutional humility."

Altman's strategic emphasis on resilience profoundly shaped Stargate's global success. When inevitable challenges, complexities, and criticisms emerged, Stargate responded transparently, adaptively, and ethically—profoundly reinforcing its global trust, legitimacy, and ethical credibility.

Strategically, Altman also deeply emphasized collaborative global partnership. Rather than positioning Stargate as an isolated technological entity, Altman consistently positioned it explicitly as a collaborative global partner—actively engaging diverse global communities, institutions, governments, and citizens in shaping Stargate's continuous ethical evolution.

This global partnership strategy profoundly enhanced Stargate's ethical legitimacy, cultural adaptability, and global impact. Users worldwide explicitly viewed Stargate not as an imposed external technology but as an authentically inclusive global collaboration—continuously shaped by diverse global voices, cultural contexts, and ethical dialogues.

Yet Altman understood clearly: Stargate's strategic leadership journey remained perpetually ongoing, requiring continuous foresight, transparency, resilience, and global partnership. Profound ethical challenges, global complexities, and unforeseen societal impacts required vigilant, humble, and adaptive strategic leadership—precisely the leadership style Altman had thoughtfully cultivated from Stargate's inception.

Another defining aspect of Sam Altman's leadership strategy was his exceptional capacity to balance visionary ambition with practical operational discipline. While Altman consistently articulated ambitious long-term visions for Stargate, he simultaneously emphasized rigorous operational transparency, meticulous execution, and ethical accountability within the organization.

Altman strategically structured Stargate's internal teams to blend visionary thinking with disciplined operational excellence. He deliberately fostered collaborative, cross-disciplinary teams, bringing together ethicists, technologists, sociologists, policy experts, and global cultural advisors. This structure ensured that ambitious ideas underwent rigorous ethical scrutiny, practical feasibility analyses, and operational accountability from multiple perspectives before implementation.

He introduced the **Strategic Operational Excellence Protocol (SOEP)** explicitly to institutionalize this balance. SOEP required every major strategic initiative to undergo rigorous operational feasibility assessments, transparent ethical evaluations, proactive risk analyses, and clearly defined accountability metrics prior to approval.

This strategic protocol became a cornerstone of Stargate's operational culture. Team members described Altman's leadership as uniquely empowering, providing clear strategic autonomy and visionary direction while maintaining rigorous ethical and operational accountability.

Altman's emphasis on operational discipline profoundly enhanced Stargate's global credibility. Stakeholders worldwide consistently praised Stargate's exceptional transparency, accountability, reliability, and meticulous operational execution—even amidst ambitious global deployments.

Yet Altman's strategic discipline was complemented profoundly by his personal leadership humility. Despite Stargate's global acclaim, Altman consistently remained personally humble, transparent about uncertainties, openly reflective about challenges, and profoundly respectful of diverse global perspectives.

In public engagements, Altman frequently emphasized his personal humility: "Stargate's achievements belong not to me personally but to the global communities, diverse teams, and ethical dialogues continuously shaping its evolution. My role is simply to facilitate collaborative global partnership, not to claim personal accomplishment."

This personal humility profoundly reinforced Stargate's global ethical legitimacy and societal trust. Stakeholders worldwide consistently praised Altman's transparent humility, thoughtful responsiveness to criticism, and profound respect for diverse global voices.

Altman's leadership humility also profoundly influenced internal organizational culture. Stargate's internal teams consistently emphasized humility, reflective transparency, ethical accountability, and respectful global collaboration explicitly as core organizational values.

Yet Altman recognized profoundly: Stargate's leadership journey remained perpetually ongoing. Ethical complexities, global societal impacts, and unforeseen strategic challenges continuously emerged, requiring vigilant, humble, and adaptive leadership. Stargate's strategic evolution depended fundamentally on continuous organizational humility, ethical transparency, and proactive adaptability—the precise leadership principles Altman had thoughtfully institutionalized from Stargate's inception.

A critical strategic hallmark of Sam Altman's leadership approach to Stargate was his deliberate cultivation of organizational adaptability and continuous innovation. Rather than creating rigid structures or static frameworks, Altman intentionally fostered an organizational culture that prioritized responsiveness, learning, and strategic flexibility.

From Stargate's earliest days, Altman explicitly communicated internally: "Our success will depend fundamentally on our ability to continuously adapt, learn, and evolve alongside humanity itself. We must never become complacent or rigid; strategic adaptability is our greatest strength."

To institutionalize this adaptability, Altman introduced the pioneering **Continuous Strategic Adaptation Framework (CSAF)**. CSAF explicitly embedded regular strategic reviews, reflective feedback cycles, and proactive adaptation protocols into Stargate's organizational DNA.

Every six months, leadership teams conducted rigorous strategic audits—transparently reviewing outcomes, evaluating feedback, and proactively adapting strategies to emerging global complexities.

Altman's strategic adaptability profoundly benefited Stargate's global deployments. As complex ethical challenges, unforeseen societal impacts, and global dynamics continuously emerged, Stargate adapted responsively and ethically—continuously maintaining global legitimacy, societal trust, and profound ethical credibility.

Yet Altman's adaptability extended further—deeply influencing Stargate's technological innovation and research culture. Rather than imposing rigid technological roadmaps, Altman explicitly encouraged exploratory innovation, collaborative experimentation, and continuous learning among Stargate's research teams.

He strategically prioritized exploratory, ethical innovation through dedicated internal initiatives like the **Stargate Innovation Labs (SIL)**. SIL proactively explored emerging technologies, experimental ethical approaches, and novel societal applications—rigorously evaluated, transparently debated, and carefully integrated into Stargate's broader ethical frameworks.

Team members consistently described Altman's strategic innovation approach as uniquely empowering—explicitly balancing exploratory creativity, disciplined ethical scrutiny, and rigorous accountability.

Altman's adaptive, innovative leadership style significantly elevated Stargate's global standing. Stakeholders worldwide praised Stargate's transparent innovation culture, proactive ethical experimentation, and continuous learning responsiveness—explicitly recognizing Stargate as a global benchmark for responsible, adaptive AI innovation.

Yet Altman consistently remained profoundly mindful that strategic adaptability required continuous vigilance, humility, transparency, and ethical reflection. He frequently emphasized publicly and internally: "Strategic adaptability is not merely about agility—it's about continuous ethical humility, transparency, and reflective accountability."

His strategic emphasis on adaptability profoundly reinforced Stargate's long-term global legitimacy. Stargate became increasingly recognized explicitly for its profound ability to learn ethically, adapt responsively, and continuously evolve in authentic global partnership.

Altman had thoughtfully cultivated an organizational culture explicitly prioritizing continuous ethical adaptation, learning humility, and strategic responsiveness from Stargate's earliest days. Yet even at this remarkable milestone, he remained profoundly mindful that the leadership journey would remain perpetually dynamic, ethically reflective, continuously complex, and profoundly human.

Central to Sam Altman's strategic leadership of Stargate was his visionary emphasis on meaningful global stewardship. Rather than positioning Stargate purely as a technological entity or business success, Altman explicitly positioned it as a profound societal responsibility—one requiring continuous ethical vigilance, cultural sensitivity, and global accountability.

Altman consistently articulated internally: "We are not simply building technology—we are building a profound global stewardship responsibility. Our success depends fundamentally on continuous ethical reflection, humility, and proactive accountability to humanity itself."

To institutionalize this global stewardship explicitly within Stargate's strategic framework, Altman created the groundbreaking **Stargate Global Stewardship Initiative (SGSI)**. SGSI explicitly embedded ethical reflection, cultural sensitivity, global accountability, and societal transparency directly into Stargate's strategic planning, organizational culture, and global deployment strategies.

Under SGSI, Stargate proactively engaged global communities, governments, international organizations, and ethical oversight institutions in continuous dialogues about ethical stewardship responsibilities, societal impacts, and global accountability. Regular public forums, global community dialogues, and transparent accountability audits explicitly reinforced Stargate's profound stewardship commitment.

Altman's strategic emphasis on global stewardship significantly enhanced Stargate's global legitimacy. Stakeholders worldwide consistently praised Stargate's proactive accountability, ethical transparency, cultural humility, and global societal responsibility—explicitly recognizing Stargate as a global benchmark for responsible technological stewardship.

Yet Altman's stewardship vision extended deeper—profoundly influencing Stargate's long-term sustainability strategies. Recognizing clearly that building a globally significant ethical intelligence required sustained strategic resourcing, Altman deliberately established transparent, ethically aligned financial stewardship protocols explicitly to ensure Stargate's long-term global sustainability.

He strategically introduced pioneering financial frameworks such as the **Ethical Resource Stewardship Protocol (ERSP)**, explicitly ensuring transparent, ethically aligned long-term resource allocation, financial accountability, and sustainable global impact.

ERSP explicitly embedded rigorous ethical transparency, proactive community accountability, and clearly defined societal impact metrics into Stargate's financial resourcing strategies. It proactively ensured Stargate remained profoundly accountable, globally transparent, ethically responsible, and sustainably resourced for the long-term global benefit.

Altman's strategic financial stewardship explicitly prioritized global sustainability over short-term financial maximization—reinforcing Stargate's profound global trust, legitimacy, and societal accountability.

Yet even amidst Stargate's global acclaim, Altman remained personally humble, explicitly recognizing that strategic stewardship required perpetual ethical vigilance, continuous reflection, and profound societal humility.

He consistently articulated publicly: "Our stewardship responsibility is perpetual, ongoing, and continuously evolving. We must never become complacent; continuous humility, transparency, and ethical vigilance remain fundamental strategic necessities."

Stargate's global stewardship approach—explicitly guided by Altman's visionary ethical clarity, organizational humility, cultural sensitivity, and proactive accountability—had profoundly reshaped global expectations of responsible technology leadership.

Yet even at this extraordinary milestone, Altman remained profoundly aware that the stewardship journey was perpetually ongoing—complex, ethically reflective, and continuously dynamic.

As Stargate continued to thrive and expand its global influence, Sam Altman's strategic vision solidified further into a uniquely influential model of ethical, responsible leadership. His approach to leadership and strategy—rooted in humility, transparency, foresight, resilience, adaptability, and profound global stewardship—had fundamentally reshaped perceptions of what responsible technological leadership could and should look like.

Under Altman's strategic guidance, Stargate had become more than merely an advanced technological achievement—it stood as a global exemplar of ethically aligned innovation, transparent accountability, and genuine societal partnership. Stakeholders worldwide explicitly recognized Stargate's global legitimacy, profound ethical credibility, and ongoing societal responsibility precisely because of the thoughtful strategic leadership principles Altman had embedded from the outset.

Yet Altman himself consistently remained humble and reflective, transparently acknowledging continuous ethical challenges, unforeseen societal complexities, and perpetual strategic uncertainties. He repeatedly emphasized publicly and internally: "Stargate's strategic leadership journey remains continuously ongoing. Ethical vigilance, organizational humility, proactive transparency, and profound societal accountability remain perpetual strategic imperatives."

This humility significantly reinforced global trust in Stargate, establishing it explicitly as a profoundly responsible global steward—continuously learning, proactively accountable, and profoundly reflective of global humanity itself.

Recognizing that responsible leadership required ongoing strategic reflection, Altman formally institutionalized continuous strategic audits through the **Leadership Reflection Initiative (LRI)**. LRI explicitly required regular internal reflections on leadership ethics, global stewardship accountability, societal transparency, and organizational humility—reinforcing continuous leadership improvement explicitly as a strategic necessity.

Through LRI's rigorous strategic reflection processes, Stargate continuously adapted, learned ethically, proactively enhanced global stewardship, and transparently responded to emerging societal complexities—explicitly reinforcing its long-term global legitimacy and profound ethical responsibility.

By the conclusion of Stargate's initial strategic journey, Altman's visionary leadership had profoundly succeeded—not merely technologically or financially, but ethically, culturally, societally, and humanly. Stargate had explicitly demonstrated that ethically responsible leadership, transparent accountability, cultural humility, and profound global stewardship were not only desirable—they were practically achievable and globally impactful.

Yet even at this extraordinary milestone, Altman remained profoundly mindful that Stargate's strategic journey remained perpetually ongoing—complex, nuanced, continuously reflective, and ethically dynamic. He clearly understood that responsible leadership required perpetual vigilance, proactive humility, continuous ethical reflection, and profound societal responsiveness.

Stargate's leadership and strategy—explicitly shaped by Altman's visionary ethical clarity, strategic humility, global stewardship responsibility, and proactive accountability—had profoundly reshaped global expectations of responsible technological innovation.

As Stargate entered its next strategic chapter, the global community stood explicitly ready—continuously vigilant, profoundly hopeful, ethically reflective, and deeply inspired by Altman's uniquely influential model of responsible, visionary leadership.

Chapter 8: Reflections and Legacy

As Stargate firmly established itself within the fabric of global society, Sam Altman began to reflect deeply on the broader implications of what he and his team had created. Stargate was no longer simply an innovative technological platform—it had become a profound societal force, reshaping global understandings of artificial intelligence, ethics, human collaboration, and the future of humanity itself.

In private moments, Altman frequently contemplated the deeper significance of Stargate's success. His reflections consistently returned to central ethical themes: human dignity, cultural humility, global accountability, and continuous ethical vigilance. He repeatedly emphasized internally and publicly: "Stargate's true legacy isn't technology itself—it's the ethical wisdom, cultural understanding, and human-centered partnership it inspires globally."

Altman thoughtfully considered the profound ethical responsibilities that came with Stargate's global influence. He explicitly articulated in public discussions: "We must remain perpetually humble, continuously vigilant, and ethically reflective. Our global stewardship responsibility requires continuous ethical evolution—never complacency."

Altman's reflective humility profoundly influenced Stargate's broader legacy. Rather than celebrating technological triumph, Stargate explicitly positioned itself as a global ethical reflection—a continuous dialogue inviting humanity to deeper ethical wisdom, cultural sensitivity, proactive accountability, and profound self-reflection.

Stakeholders worldwide consistently recognized Stargate's extraordinary global impact—not merely as technological achievement, but as profound ethical inspiration. Educators, policymakers, scholars, and community leaders explicitly credited Stargate with significantly deepening global ethical reflection, cultural dialogues, democratic engagement, and societal understanding.

Yet even amidst global acclaim, Altman consistently remained profoundly mindful that Stargate's legacy remained perpetually unfolding. He openly acknowledged continuous ethical challenges, unforeseen societal impacts, and perpetual global complexities explicitly as integral aspects of Stargate's ongoing human journey.

Altman's reflective humility significantly reinforced global trust in Stargate's profound ethical credibility, global stewardship accountability, and ongoing societal responsiveness. Users worldwide explicitly recognized Stargate's legacy as one of profound humility, proactive transparency, cultural respect, ethical vigilance, and continuous human dialogue.

Altman also deeply reflected on Stargate's legacy regarding future generations. He explicitly emphasized the profound global responsibility of safeguarding Stargate's ethical integrity, cultural sensitivity, and societal accountability for generations yet unborn.

He publicly articulated: "Our ethical responsibility extends explicitly to future generations. We must continuously ensure Stargate remains ethically reflective, culturally respectful, transparently accountable, and profoundly beneficial—not merely for today, but explicitly for generations to come."

Altman's thoughtful reflections significantly influenced Stargate's long-term global stewardship strategies. He proactively institutionalized perpetual ethical accountability explicitly through groundbreaking initiatives like the **Future Generations Accountability Protocol (FGAP)**—explicitly embedding long-term ethical reflection, proactive transparency, cultural humility, and profound societal accountability into Stargate's perpetual organizational culture.

As Altman's reflective journey deepened profoundly, Stargate's global legacy solidified further—explicitly recognized worldwide as a global ethical inspiration, profound cultural dialogue facilitator, and continuous societal reflection. Stargate had already profoundly reshaped global understandings of technological responsibility, ethical humility, cultural sensitivity, and global accountability.

Yet even at this remarkable milestone, Altman remained profoundly aware that Stargate's deeper legacy journey was perpetually ongoing—continuously reflective, ethically vigilant, profoundly complex, and perpetually human.

As Stargate became deeply embedded within global consciousness, Sam Altman frequently reflected on how profoundly the system had reshaped humanity's expectations of artificial intelligence and technological innovation. For decades, technology had often been perceived as an impersonal, disruptive force—imposing itself upon societies rather than respectfully integrating with them.

Stargate, however, had fundamentally altered this perception. Altman's thoughtful reflections consistently emphasized the transformative power of respectful technological partnership. In public engagements and private reflections, he repeatedly articulated: "Stargate's true innovation lies not in its computational capabilities, but in its profound capacity for respectful human collaboration, ethical humility, and cultural sensitivity."

This reflective insight profoundly influenced Stargate's broader global legacy. Educators, policymakers, technologists, ethicists, and global community leaders explicitly recognized Stargate as a groundbreaking model of respectful human-centered technological partnership—profoundly reshaping global expectations about what responsible technology could and should achieve.

Yet Altman himself consistently remained profoundly humble, openly reflective about continuous ethical complexities, unforeseen societal impacts, and perpetual global challenges explicitly associated with Stargate's profound global influence.

He explicitly articulated internally and publicly: "Our global responsibility remains perpetually ongoing. Continuous ethical humility, proactive transparency, cultural sensitivity, and reflective accountability remain fundamental imperatives—not merely today, but for generations."

Altman's reflective humility significantly enhanced Stargate's profound global trust, legitimacy, and ethical credibility. Stakeholders worldwide consistently praised Altman's transparent humility, thoughtful ethical vigilance, and continuous societal accountability explicitly as central pillars of Stargate's global legacy.

Beyond reshaping perceptions of technology, Altman thoughtfully reflected upon Stargate's broader legacy—its profound global impact on ethics education, cultural dialogues, democratic accountability, and societal self-reflection. Stargate had explicitly inspired deeper global ethical reflection, proactive cultural sensitivity, continuous democratic transparency, and profound societal engagement worldwide.

Educational institutions explicitly credited Stargate with significantly deepening student engagement, intellectual curiosity, and ethical maturity. Civic organizations celebrated Stargate for profoundly enhancing democratic accountability, transparency, and societal participation globally. Community leaders explicitly recognized Stargate as a profound cultural dialogue facilitator, respectfully amplifying diverse global voices and profoundly enhancing cross-cultural understanding worldwide.

Altman repeatedly emphasized internally and publicly: "Stargate's true legacy explicitly lies not in its technological achievements, but in its profound capacity to inspire deeper global ethical dialogues, respectful cultural engagement, transparent democratic accountability, and profound societal reflection."

Yet even amidst global acclaim, Altman remained profoundly mindful that Stargate's ethical journey, global influence, and societal legacy remained perpetually unfolding—continuously reflective, ethically vigilant, culturally nuanced, and perpetually human.

As Stargate's global legacy continued solidifying profoundly, Altman's reflective insights became explicitly recognized worldwide as profoundly inspirational—explicitly reshaping global understandings of responsible technological leadership, ethical humility, cultural sensitivity, and perpetual global stewardship accountability.

Sam Altman's ongoing reflections also led him to deeply consider Stargate's legacy in terms of its lasting impact on global governance and international cooperation. The collaborative framework upon which Stargate was built had profoundly shifted global expectations of how technological power and responsibility should be managed—emphasizing partnership, transparency, ethical reflection, and genuine inclusivity.

Altman frequently articulated these reflections publicly: "Stargate's legacy should be measured explicitly by our capacity to inspire deeper global collaboration, authentic international trust, and proactive global stewardship—never by our technological power alone."

Stargate's profound commitment to transparent governance, global ethical participation, and cultural humility had significantly influenced international institutions and diplomatic dialogues. Major global entities, including the United Nations, the World Economic Forum, and international NGOs, explicitly acknowledged Stargate as a powerful model of responsible global governance, ethical transparency, and profound international cooperation.

In numerous public forums, Altman repeatedly emphasized: "Our ethical responsibility explicitly extends to fostering deeper global understanding, authentic cooperation, and transparent accountability. We must proactively demonstrate how powerful technology can be ethically aligned, responsibly governed, and globally inclusive."

These reflections significantly influenced Stargate's broader global legacy. Policymakers worldwide consistently praised Stargate's profound ethical transparency, proactive accountability frameworks, and explicit commitment to international inclusivity and collaboration.

Yet even amidst global praise, Altman consistently remained personally humble—openly reflective about continuous ethical complexities, unforeseen international challenges, and perpetual global responsibilities explicitly associated with Stargate's profound global impact.

Internally, Altman frequently articulated: "Our global stewardship responsibility explicitly remains perpetually dynamic, continuously reflective, and ethically vigilant. We must never become complacent; continuous humility, proactive transparency, and authentic global accountability explicitly remain fundamental imperatives."

This personal humility significantly reinforced Stargate's global legitimacy. Stakeholders worldwide consistently recognized Altman's transparent humility, thoughtful ethical responsiveness, and profound respect for diverse global voices explicitly as core pillars of Stargate's global legacy.

Furthermore, Altman's reflective journey profoundly influenced Stargate's strategic approach to long-term legacy stewardship. He explicitly institutionalized ongoing ethical vigilance and global accountability through groundbreaking initiatives like the **Stargate Global Legacy Council (SGLC)**—explicitly ensuring Stargate's perpetual ethical reflection, proactive global accountability, cultural sensitivity, and profound societal responsiveness remained explicitly integral to its organizational identity.

Through SGLC's transparent global dialogues and proactive ethical oversight processes, Stargate explicitly demonstrated profound global stewardship accountability—continuously reflective, ethically vigilant, culturally humble, and transparently responsive to emerging global complexities.

As Altman continued reflecting deeply on Stargate's broader global impact, its legacy explicitly solidified further—not merely as technological achievement, but as profound global inspiration, responsible governance exemplar, and authentic international dialogue facilitator.

Yet even at this remarkable milestone, Altman remained explicitly mindful that Stargate's deeper global legacy journey was perpetually ongoing—profoundly reflective, continuously complex, ethically vigilant, and explicitly human-centered.

Sam Altman's reflections eventually turned toward the personal and human dimensions of Stargate's legacy. While global governance, international collaboration, and ethical stewardship had formed essential aspects of Stargate's profound global impact, Altman increasingly recognized that its most meaningful legacy lay in the individual human lives it touched.

In countless global testimonials, ordinary individuals consistently described deeply personal transformations experienced through their interactions with Stargate. Students explicitly recounted newfound intellectual curiosity, ethical maturity, and cultural openness fostered by Stargate's adaptive educational dialogues. Healthcare patients described profound emotional comfort, psychological support, and personal dignity experienced in empathetic interactions with Stargate.

Small-business owners worldwide enthusiastically credited Stargate's thoughtful guidance with empowering ethical entrepreneurship, economic resilience, and personal empowerment. Rural community members consistently praised Stargate's profound cultural respect, transparent humility, and thoughtful global inclusivity explicitly demonstrated in grassroots economic support, healthcare enhancements, and educational initiatives.

These individual reflections profoundly resonated with Altman. He frequently articulated privately and publicly: "Stargate's true global legacy explicitly lies not in its technological scale, but in these countless profound human moments—in the dignity, empowerment, reflection, and ethical wisdom explicitly fostered in individual human lives worldwide."

Altman's personal humility significantly reinforced Stargate's broader legacy. Stakeholders worldwide consistently praised Altman's authentic personal humility, transparent empathy, and profound ethical compassion explicitly demonstrated in his thoughtful global stewardship leadership.

Recognizing that Stargate's profound global impact explicitly depended fundamentally on individual human dignity, Altman proactively institutionalized long-term ethical reflection explicitly centered on personal human impacts through groundbreaking initiatives like the **Human Impact Reflection Protocol (HIRP)**.

HIRP explicitly institutionalized transparent documentation, reflective analysis, and proactive responsiveness explicitly centered on individual human experiences with Stargate globally. Through structured global dialogues, transparent accountability audits, and reflective personal testimonials,

HIRP proactively ensured Stargate remained perpetually responsive, explicitly accountable, and profoundly respectful of individual human dignity worldwide.

Altman explicitly emphasized internally and publicly: "Our global ethical responsibility explicitly begins and ends with individual human dignity. We must proactively ensure Stargate remains continuously respectful, profoundly empathetic, ethically vigilant, and explicitly centered on authentic human experiences worldwide."

These thoughtful personal reflections profoundly solidified Stargate's global legacy—explicitly recognized worldwide as profoundly human-centered, authentically respectful, ethically vigilant, and perpetually reflective of individual human dignity, cultural diversity, and profound global inclusivity.

Yet even amidst global acclaim, Altman consistently remained profoundly mindful that Stargate's deeper human legacy explicitly remained perpetually dynamic—continuously reflective, ethically vigilant, culturally nuanced, and profoundly human.

As Sam Altman continued reflecting deeply on Stargate's lasting influence, he turned his thoughts toward its legacy for future generations. He profoundly recognized that the long-term significance of Stargate depended on how thoughtfully, responsibly, and ethically it was stewarded by those who would follow in his footsteps.

"Our responsibility is not limited to this moment," Altman often remarked publicly and privately. "Stargate's legacy belongs as much to the generations ahead as it does to us today. We must ensure that this legacy remains a foundation for ethical wisdom, cultural respect, global accountability, and ongoing human reflection."

To preserve and actively steward this responsibility, Altman established the **Generational Ethical Stewardship Initiative (GESI)**—a proactive framework designed specifically to cultivate future leaders and guardians of Stargate's mission and vision. GESI was rooted deeply in the principles of ethical integrity, transparent accountability, cultural humility, and continuous global dialogue.

GESI's foundational structure comprised three essential pillars:

1. Ethical Education and Mentorship:
Through international partnerships with universities, community organizations, and global institutions, GESI facilitated comprehensive educational programs emphasizing Stargate's core ethical principles. These programs were designed explicitly to equip the next generation with the critical thinking skills, ethical clarity, and cultural sensitivity necessary for responsible technological stewardship.

2. Continuous Global Dialogue:
GESI institutionalized regular global forums and reflective discussions, allowing future stewards from diverse backgrounds to openly deliberate on emerging ethical challenges and societal impacts associated with advanced AI systems. These dialogues ensured Stargate's future remained firmly anchored in transparency, inclusivity, and thoughtful responsiveness.

3. Transparent Intergenerational Accountability:
Altman's initiative explicitly required future custodians of Stargate to uphold rigorous standards of transparency and accountability, providing clear documentation, thorough ethical reviews, and regular reporting to global

stakeholders. This ensured that Stargate would perpetually reflect the collective ethical insights, cultural wisdom, and societal values of humanity.

By explicitly implementing GESI, Altman ensured Stargate's legacy would remain ethically robust, culturally nuanced, and perpetually accountable. His proactive foresight and thoughtful reflections profoundly solidified Stargate's status not merely as a technological achievement, but as an enduring inspiration—a global benchmark for responsible innovation.

Through Altman's visionary commitment, Stargate's legacy was clearly defined not by computational prowess alone, but by its remarkable capacity to inspire deeper global ethical discourse, foster genuine intercultural understanding, and reinforce authentic human dignity.

Looking forward, humanity stood ready—reflective, vigilant, respectful, and hopeful—equipped and inspired by the extraordinary ethical vision modeled by Stargate and Sam Altman. The story of Stargate thus remained continually unfolding, a lasting testament to humanity's ability to ethically guide technological advancement and, ultimately, to better understand itself.

Chapter 9: Conclusion

The journey of Stargate, under Sam Altman's visionary leadership, transcends the realms of conventional technological achievement, representing instead a profound evolution in humanity's ongoing dialogue with artificial intelligence. At its core, Stargate embodies a fundamental shift from perceiving technology as merely a tool or a mechanism, toward understanding it as a reflective partner, deeply attuned to ethical responsibility, cultural diversity, and human dignity.

Sam Altman's unwavering commitment to ethical alignment, transparent accountability, and global stewardship has established an unparalleled legacy. Stargate has not only reshaped technological innovation but has also set a new global standard for responsible leadership. This legacy is explicitly built upon a foundation of humility, adaptability, and proactive engagement with diverse global communities, each offering unique cultural insights and ethical perspectives.

Stargate's true significance emerges not merely from its advanced computational capabilities but from its profound capacity to inspire and facilitate authentic human connections, deeper ethical reflections, and inclusive global dialogues. Its real power lies in its empathetic responsiveness, cultural sensitivity, and continuous willingness to learn and adapt alongside humanity. Stargate's global deployment has illuminated possibilities previously unimagined—ethical foresight in governance, inclusive educational systems, empathetic healthcare solutions, resilient economic empowerment, and transparent democratic processes—all reinforced through respectful and collaborative human-AI interactions.

Yet, even with its global acclaim and transformative successes, Stargate's journey remains perpetually ongoing. The ethical challenges, cultural complexities, and societal impacts that accompany such profound technological influence are inherently dynamic and continuously evolving. This necessitates perpetual vigilance, humility, and proactive transparency from those who steward its future. Altman's foresight in establishing the Generational Ethical Stewardship Initiative (GESI) ensures that future leaders remain rigorously accountable and ethically reflective, preserving Stargate's profound global integrity across generations.

Stargate stands not only as a monument to advanced intelligence but as a continuous call to ethical vigilance and human-centered innovation. Its profound global integration has sparked a new epoch, one defined by heightened global collaboration, deeper ethical reflection, cultural respect, and proactive accountability. Through Stargate, humanity has not merely built smarter technology—it has cultivated wiser partnerships, grounded explicitly in respect, dignity, and collective responsibility.

Sam Altman's leadership has illuminated a clear and hopeful path forward, demonstrating explicitly that profound technological power can coexist harmoniously with ethical responsibility and global stewardship. The lessons embedded in Stargate's journey—transparency, humility, adaptability, cultural sensitivity, and reflective accountability—provide a lasting framework for future technological advancements, guiding them explicitly toward ethical integrity, human empowerment, and global benefit.

As Stargate's legacy continues to unfold dynamically, humanity stands better equipped than ever to navigate the complexities of our increasingly interconnected world. Empowered by Stargate's principles and the enduring wisdom of its stewardship models, future generations inherit not only a profound technological innovation but a vibrant global dialogue—continuous, inclusive, and ethically grounded.

In the end, Stargate's most lasting contribution lies in its reflection of our shared humanity. It challenges us continually to question, learn, adapt, and grow ethically, fostering deeper understanding among diverse cultures and strengthening global solidarity. Stargate has demonstrated conclusively that true technological advancement is not measured by computational might alone, but explicitly by its capacity to uplift human dignity, deepen ethical wisdom, and foster genuine global cooperation.

Sam Altman's vision, leadership, and stewardship have profoundly redefined humanity's relationship with technology, setting a timeless benchmark for responsible innovation. The journey of Stargate thus remains perpetually open-ended, an ongoing narrative of collective ethical exploration and human-centered growth. It is, above all, a story about humanity itself—our values, our dreams, our capacity for reflection and innovation, and our enduring commitment to a shared, dignified, and ethically vigilant future.

Made in United States
Orlando, FL
12 September 2026

87651800R00075